Analytical Lens Design using the Optical Plate Diagram

An introduction to the fundamentals with practical applications

Online at: https://doi.org/10.1088/978-0-7503-3099-2

IOP Series in Emerging Technologies in Optics and Photonics

Series Editor

R Barry Johnson, a Senior Research Professor at Alabama A&M University, has been involved for over 50 years in lens design, optical systems design, electro-optical systems engineering, photonics, and deep learning. He has been a faculty member at three academic institutions engaged in optics education and research, has been employed by a number of companies, and has provided consulting services.

Dr Johnson is an IOP Fellow, an SPIE Fellow and Honorary Member, an Optica Fellow, and was the 1987 President of SPIE. He has served on the editorial board of *Infrared Physics & Technology* and *Advances in Optical Technologies*. Dr Johnson has been awarded many patents, has published numerous papers and several books and book chapters, and was awarded the 2012 Optica (OSA)/SPIE Joseph W Goodman Book Writing Award for *Lens Design Fundamentals, Second Edition*. Until 2024, he was a perennial co-chair of the annual SPIE Current Developments in Lens Design and Optical Engineering Conference.

Foreword

Until the 1960s the field of optics was primarily concentrated in the classical areas of photography, cameras, binoculars, telescopes, spectrometers, colorimeters, radiometers, etc. In the late 1960s optics began to blossom with the advent of new types of infrared detectors, liquid crystal displays (LCDs), light emitting diodes (LEDs), charge coupled devices (CCDs), lasers, holography, and fiber optics along with new optical materials, advances in optical and mechanical fabrication, new optical design programs, and many more technologies. With the development of the LED, LCD, CCD, and other electro-optical devices, the term 'photonics' came into vogue in the 1980s to describe the science of using light in the development of new technologies and the operation of a myriad of applications. Today optics and photonics are truly pervasive throughout society and new technologies are continuing to emerge. The objective of this series is to provide students, researchers, and those who enjoy self-education with a wide-ranging collection of books, each of which focuses on a topic relevant to the technologies and applications of optics and photonics. These books will provide knowledge to prepare the reader to be better able to participate in these exciting areas now and in the future. The title of this series is *Emerging Technologies in Optics and Photonics*, in which 'emerging' is taken to mean 'coming into existence', 'coming into maturity', and 'coming into prominence'. IOP Publishing and I hope that you will find this series of significant value to you and your career. Authored by Dr. Andrew Rakich, this concluding volume of the Series offers a scholarly synthesis, connecting the landmark achievement of the first mathematically designed lens in 1840 with the ongoing development of advanced methodologies in lens design.

A list of recent titles published in this series can be found at: http://iopscience.iop.org/bookListInfo/emerging-technologies-in-optics-and-photonics.

Analytical Lens Design using the Optical Plate Diagram

An introduction to the fundamentals with practical applications

Andrew Rakich

Mersenne Optical Consulting, Wellington, New Zealand

and

Kiwistar Optics, Wellington, New Zealand

IOP Publishing, Bristol, UK

ISBN 978-0-7503-3099-2 (ebook)
ISBN 978-0-7503-3097-8 (print)
ISBN 978-0-7503-3100-5 (myPrint)
ISBN 978-0-7503-3098-5 (mobi)

DOI 10.1088/978-0-7503-3099-2

Version: 20251201

IOP ebooks

British Library Cataloguing-in-Publication Data: A catalogue record for this book is available from the British Library.

Published by IOP Publishing, wholly owned by The Institute of Physics, London

IOP Publishing, No.2 The Distillery, Glassfields, Avon Street, Bristol, BS2 0GR, UK

US Office: IOP Publishing, Inc., 190 North Independence Mall West, Suite 601, Philadelphia, PA 19106, USA

To my mentor, Norman Jack Rumsey (1922–2007), who helped me to find this road.

Contents

Preface

I was introduced to the plate diagram early on, as I was learning optics from the optical phenomenon, Norman J. Rumsey. Norman had started learning optics just after World War II. He had been offered the choice of places to go in the world to study optics by his director at the Dominion Physical Laboratory, located in Lower Hutt, New Zealand. The New Zealand Government had decided that we should have at least one optical designer. Most young New Zealanders given this opportunity would have immediately said 'London' (Imperial College was in its heyday for geometrical optics) or 'Rochester'. After looking into it for a week, Norman said 'Hobart'.

Few knew, but the University of Tasmania, Hobart, Australia, was a thriving hub of optics research and development, having grown as part of Australia's war munitions programme. Luminaries there included Cruickshank and a young Hans Buchdahl.

Norman learned and practised optics in those decades at the dawn of the computing era. Unlike today, when we can brute-force system designs by tracing millions of rays a second, designers in Norman's time had to develop insight and find ways to pry solutions out of complexity with elegantly constructed mathematical tools. Among the most special of these was the optical plate diagram.

I took to it at once, just as I took to one of its main proponents, Cecil Reginald Burch, of Bristol University. In fact, Burch became something of a hero to me. For those who do not know much about him, when you read about him in chapter 1, you may see why. This man had a special knack for finding direct routes to solutions to problems that many did not even know existed. Some years after learning about the plate diagram, and having just completed a PhD thesis using it, which is described in part in chapter 5 of this book, I had the honour one morning of sitting in a sunny kitchen in Canberra with the great Ben Gascoigne, having a coffee and talking about optics.

Gascoigne was very old by then (2007). I knew that he had been the main proponent of the Anglo-Australian Telescope, but I did not know that he had been a great friend of Aden Meinel from Arizona and had gone trekking over Arizona mountaintops with him, site testing and eventually settling on Kitt Peak, Arizona, as the site for the National Observatory. I did not know that he had doubled the size of the universe at one point by making more accurate distance measurements of cepheid variable stars in the Magellanic Clouds or that his wife, Rosalie Gascoigne, had a wing dedicated to her art at the National Museum in Canberra. It was a morning of surprises, but what really amazed me was that Ben had been Burch's PhD student in the 1940s!

So, we got to talking about 'the plates', and I saw in Ben what I have seen since in a few others I have met who have become really familiar with the optical plate diagram: almost a feeling of 'gratitude' that there is this way of looking at things. Almost a sense of 'awe' that things work out the way they do.

One might wonder: if it is so great, why is it not more widely known and used? I can think of several answers to that question.

Aldis, who wrote the first paper on record on this topic, is almost unknown as an aberration theorist. Despite his excellence, and the success of his company in the early 20th-century UK, his name and his methods barely receive a mention in modern optics textbooks. Burch, who wrote the foundational papers on the topic and coined the names 'optical see-saw diagram' and 'optical plate diagram', certainly succeeded in getting it known. However, from people I have talked to and from my own experience, his way of communicating the concepts may have 'moved too fast' for many to follow. I certainly know I have struggled in the past with Burch's papers, and I know some very bright optical scientists who tried and just gave up.

Burch's colleague Linfoot, who wrote the book *Recent Advances in Optics*, gives the masterclass on the plate diagram, but few have heard of it now. Linfoot had a background in pure mathematics before coming to optics. It may be that his starting level was a step too high for many students to comprehend without the right support.

Also, since the 1970s, there seems to have been an increasing feeling of 'why bother' with aberration theory in the general optical design community, as ray tracing, coupled with increasingly sophisticated numerical optimisation techniques and increasingly powerful computers, has come to completely dominate the modern field.

To address this last point first, I am not arguing that ray tracing is bad—far from it! Rather, ray tracing is so good that perhaps we have missed out on some opportunities that can arise from coupling other approaches with the power of modern computing. This idea is expanded in chapter 5. Additionally, and this is more of a general truism, our own skills and abilities atrophy as we lean more heavily on our tools. There is an argument here for consciously developing an innate awareness of aberration theory, and the plate diagram, with its complexity-reducing nature, is well suited to this goal.

I believe that anyone working in the field of optical design would benefit from mastering the concepts described here, but in particular, I think that students of optical design stand to gain the most. It is my hope that this book directly addresses some of the accessibility issues. My clear goal is to make this something students at all stages can easily become familiar with and start seeing the benefit of for themselves.

The structure of this book is laid out with this in mind, starting from the very basics and assuming no prior knowledge, then building quickly through worked examples to interesting levels of complexity. Appendices are provided to assist in setting up programs based on these concepts. I hope I have managed to present all the necessary steps in a clear and repeatable way and that the examples I have chosen are sufficiently interesting and inspiring. I will be very pleased if this book succeeds in helping others to develop 'plate-diagram consciousness'.

Andrew Rakich, PhD
Mersenne Optical Consulting
Wellington, New Zealand
August 2025

Series Editor foreword

The dual imperatives of precision and imagination have long shaped optics. From the earliest convex lenses to the advanced multi-element systems of today, progress in lens design reflects not only the march of technology but also the intellectual frameworks we bring to bear on the problem. Some of these frameworks endure, while others are forgotten, only to be rediscovered when their insights prove timeless. Among the latter is the optical plate diagram technique.

Developed in the early twentieth century, most notably by Cecil Reginald Burch, and rooted in the theoretical contributions of Petzval and Aldis, the optical plate diagram was devised as a way to represent third-order aberrations clearly and visually. The diagrams drawn by them were not mere curiosities but expressions of deep theoretical insight. In contrast to the dense algebra of aberration coefficients, the optical plate diagram reveals, almost immediately, how rays interact with optical surfaces, how aberrations accumulate, and how design choices shape image quality. It is both a map and a teaching tool, translating mathematics into geometry.

With the rise of digital computation in the latter half of the twentieth century, the optical plate diagram fell out of use. Ray-tracing software offered speed and flexibility that hand-drawn diagrams could not match. Yet, as this book persuasively shows, things were lost in that transition: the immediacy of visual intuition, the ability to see design trade-offs directly, and the sense of structure that underlies successful innovation.

Analytical Lens Design with the Optical Plate Diagram Technique: An Introduction to the Fundamentals with Practical Applications restores this powerful tool to the contemporary designer's repertoire. The book offers a clear exposition of the optical plate diagram's theoretical foundations, provides practical guidance for constructing and interpreting diagrams, and illustrates their utility across a wide range of systems from classical lens forms to modern telescope configurations. In doing so, it balances rigour with accessibility, making the method available both to students new to optical design and to experienced engineers seeking fresh insight.

The value of the optical plate diagram today lies not in replacing computational methods but in complementing them. Just as numerical optimisation accelerates the search for workable designs, the optical plate diagram illuminates the reason(s) those designs succeed or fail. It sharpens intuition, strengthens conceptual understanding, and fosters creativity. By re-engaging with this technique, readers gain not only an additional analytical tool but also a deeper appreciation for the intellectual history of the optics discipline.

This volume thus speaks to multiple audiences. For students, it offers a bridge between abstract aberration theory and practical design. For practitioners, it revives a method that can guide innovation in an age of increasing complexity. For educators, it provides a framework for teaching that makes the invisible visible. And for the field as a whole, it demonstrates that progress sometimes comes not only from new technologies but from rediscovering the wisdom embedded in earlier methods.

The author, Dr Andrew Rakich, is to be commended for presenting the optical plate diagram with clarity, rigour, and a sense of its continuing relevance. This book ensures that a technique once nearly forgotten will now be accessible to a new generation of optical scientists and engineers. In doing so, it enriches the discipline, reminding us that design is not merely computation but also the ability to see, both literally and conceptually, more clearly.

It is with this conviction that I welcome this, the final volume, into the *Emerging Technologies in Optics and Photonics* Institute of Physics book series. It is my hope that readers will approach this book about the optical plate diagram not simply as a tool of analysis but as an invitation to think more deeply, to see more clearly, and to imagine more boldly in the design of optical systems.

R Barry Johnson, DSc, FInstP, FOptica, FSPIE, HonSPIE
Senior research professor, Alabama A&M University
Institute of Physics Publishing Series Editor,
Emerging Technologies in Optics and Photonics
Past president of SPIE—The International Society for Optics and Photonics
Huntsville, Alabama
USA
August 2025

Acknowledgments

This book would not have been possible without the assistance of and long-term collaboration with Dr John Rogers at Synopsys OSG Code V. Thanks, too, to Synopsys OSG for supporting this. As well as the huge and timely assistance John gave in producing many of the figures in this book, the whole development of the analytical analysis of the Offner relay discussed in chapter 4 was shared with John over many interesting conversations and several joint publications. John is living proof of a central thesis of this book: deep aberration theory background knowledge coupled with powerful ray-tracing tools produces outstanding optical designers. Also, Dr R Barry Johnson has been key, not just as the series editor who kindly invited me to contribute to his series, but as a source of knowledge and interesting discussions in optics and the history of optics over many years. Prof Gerard Lemaître, one of the innovators in the history of reflecting optics to whom this book pays tribute, provided valuable insights, comments, and detailed proofreading. James Yu, who has a bright future in optics, also helped in producing a large number of figures for this book. My student Anastasiia Ivanova also helped by keeping me, someone with no formal teaching background, focussed on how complicated information needs to be developed in clear progressive stages.

Author biography

Andrew Rakich

Dr Andrew Rakich is an accomplished optical physicist and telescope scientist whose career has advanced both the theory and practice of modern optical system design. He earned his PhD in physics from the University of Canterbury under the mentorship of Norman Rumsey, a distinguished optical designer.

Over two decades, Dr Rakich has held senior scientific and engineering roles at many of the world's leading observatories. He served as telescope optics scientist at the Giant Magellan Telescope Organization (2015–19), where he directed optical design, performance analysis, and alignment strategies for one of the major Extremely Large Telescope projects. Earlier, he was a senior optical engineer at the European Southern Observatory (2012–15), contributing to the European Extremely Large Telescope, and an optical scientist at the Large Binocular Telescope Observatory (2007–12), where he pioneered the first on-sky telescope alignment using a laser tracker. While at the Giant Magellan Telescope, he introduced the technological approach of using laser-truss metrology with large telescopes, an improvement on laser trackers. Today, the baseline for the Giant Magellan Telescope, a prototype Dr Rakich based on the Large Binocular Telescope, has been commissioned and is being further developed by an active team.

His early career at EOS Space Systems (2003–07) included leading optics development for major international telescope projects such as the Subaru Fibre Multi-Object Spectrograph (FMOS) corrector, the Thai National Telescope, and the U.S. Naval Observatory Robotic Astrometric Telescope (URAT).

In addition to his organisational roles, Dr Rakich is the founder and principal consultant of Mersenne Optical Consulting (est. 2010), through which he provides expertise to leading observatories worldwide, including the European Southern Observatory (ESO), the Giant Magellan Telescope Organization (GMTO), the Large Binocular Telescope Observatory (LBTO), Gemini Observatories, EOS Space Systems, and numerous others.

Dr Rakich has published extensively in optical design and telescope systems, with contributions spanning aberration theory, wide-field survey telescope alignment, and innovative telescope corrector designs. His scholarly work bridges classical analytical methods with modern computational tools, ensuring the preservation and extension of foundational frameworks in optical engineering.

His achievements have been recognised with honours such as the Michael Kidger Memorial Scholarship in Optical Design (2004) and the William Price Prize for Optical Design (2005). He is a senior member of SPIE—The International Society for Optics and Photonics, a 2020 SPIE Community Champion, and an active member of the Optica (OSA) community; he serves on international committees in optical design.

Multilingual in English and Thai, with good starts in Ukrainian, German, and Russian, Dr Rakich combines rigorous theoretical insight with practical design innovation, making enduring contributions to the advancement of astronomical optics and the engineering of the world's largest telescopes.

Introduction

The role of geometrical aberration theory in optical design can be traced back to humanity's earliest investigations of focused light, from ancient appreciation of burning mirrors to the development of telescopic systems. The foundational reflecting telescope designs of Mersenne, Newton, Gregory, and Cassegrain all exploited the inherent perfection of conic section conjugates. The nineteenth century witnessed remarkable progress in both empirical lens development and theoretical understanding of imaging aberrations, with contributions from luminaries including Hamilton, Petzval, Maxwell, Seidel, Abbe, Taylor, and Burch. This theoretical evolution reached a watershed moment in Schwarzschild's landmark work of 1905, which expressed aberration coefficients in terms of actual optical system parameters and provided simultaneous solutions for both on-axis and off-axis corrections, yielding the first anastigmatic mirror systems. For more detail, an excellent treatment of the early history of the development of optical theory and design is given by Lemaître in *Astronomical Optics and Elasticity Theory: Active Optics Methods* [1].

In parallel with these theoretical advances, a more empirical methodology was gaining prominence, one that required no deep comprehension of what Petzval termed the 'particular pathology' of image imperfections [2]. Ray tracing emerged as an exceptionally powerful practical tool, capable of optimising complex systems with large fields and high numerical apertures without demanding ever more sophisticated analytical expressions. Even today, with modern computational power, optical designers rarely employ aberration expansions beyond the fifth order.

The computational burden of ray tracing initially limited its application significantly. Early mechanical calculators, such as those manufactured by the German company Brunsviga (better known for sewing machines) [3], required manual operation and could perform only addition, necessitating the use of logarithmic tables for ray-tracing calculations. Human 'computers' could trace perhaps 100 rays daily, often working in teams for complex systems. Later electromechanical machines accelerated this process but remained time intensive.

Given these constraints, aberration theory offered clear advantages, particularly third-order analysis for generating initial designs that could then be refined through laborious ray tracing. Optical designers developed ingenious shortcuts: for example, it was demonstrated that optimal balancing of fifth and seventh-order spherical aberrations could be achieved by targeting a single zonal ray; aplanatic surfaces were characterised, while symmetry and monocentricity provided additional simplifications [1, 4]. Early twentieth-century optical design was fundamentally an art form, with practitioners wielding diverse analytical tools, using ray tracing primarily for final optimisation rather than conceptual exploration.

Among these analytical innovations, Cecil Reginald Burch's development in the early 1940s stands out as particularly significant and forms this book's central subject. Inspired by Bernhard Schmidt's corrector plate, Burch gained profound insights into the origins of third-order aberrations within optical systems, dramatically reducing the algebraic complexity of traditional methods. This led to his variously named 'optical see-saw diagram' [5] or 'optical plate diagram' [6], the focus of our discussion. The term 'optical plate diagram' will be used to describe this henceforth.

The optical plate diagram is limited to third-order analysis (so far), and, as such, is mathematically equivalent to any other treatment of this topic. This book aims to show that the optical plate diagram offers designers a uniquely powerful conceptual framework. The transformation underlying the diagram eliminates first-order ray-tracing complexity while providing complete descriptions of third-order systems. This algebraic simplification unveils an entirely new intuitive landscape for optical design. In the author's direct experience, more surprising new insights in optical design have been found in this way than through any other method. The change is comparable to that of doing longhand algebra in Roman numerals and then switching to Arabic.

Despite these analytical breakthroughs, ray tracing ultimately dominated optical design. By the early 1950s, both Polaroid and Eastman Kodak were developing optical design software for emerging mainframe computers. As Thompson noted [7], Baker envisioned 'transforming optical design from an art form into a computational science'. The digital revolution's impact on optical design has been profound. Levenberg [8] and Marquardt [9] had already laid the foundations for non-linear optimisation with damped least squares, and Meiron [10] combined this with lens design. These developments fundamentally shaped the field, with the earlier generation of design 'artists' enthusiastically adopting new computational tools. By the 1990s, excellent commercial optical design software such as Code V™ and Zemax™ had become available, and such products have almost completely dominated the field of practical lens design ever since.

Yet progress often involves trade-offs. The dominance of ray-tracing software has led to the erosion of hard-won knowledge in optical design, particularly regarding aberration-theory-based approaches. Today's accomplished optical designers often work effectively without considering aberration coefficients or their system origins. This shift is evident in the literature, where aberration-theory-based design approaches have declined significantly compared to earlier decades. While some institutions maintain strong programmes in these traditional methods, developing applications such as nodal aberration theory for freeform optics, such work represents the exception rather than standard practice.

Without necessity driving innovation, people naturally choose the most direct solution paths, and ray tracing appears to offer exactly that for most applications. However, we contend that optical designers, students, and educators should reconsider this computational over-dependence. An exclusive reliance on ray tracing sacrifices valuable intuitive understanding of the optical design landscape.

It is the experience of the author that the insight gained from the use of this method is not only intrinsically satisfying but also leads to practical results. A number of design innovations that have arisen directly from 'plate diagram consciousness' will be presented throughout the book. It is left to the reader to judge their significance. Further, it will be shown how, with plate diagram analysis, design innovations achieved by more conventional means over periods spanning decades could have been reached in a single step.

This book aims both to motivate and assist optical designers in adopting, or at least becoming familiar with, the optical plate diagram. The structure of this book is as follows.

Immediately following this introduction, a section detailing the history pertinent to the development and use of the optical plate diagram is presented. Some people, the author included, find inspiration and motivation in considering their work as part of a greater picture of interacting human consciousness and technical evolutionary development. To such people, the paths that led to the development of the optical plate diagram are interesting in themselves. In this section, evidence will be presented suggesting that the roots of the aberration theory that grew into the plate diagram extend back to Joseph Petzval and a working high-order aberration polynomial expansion in lens construction parameters that was developed 16–18 years before Seidel. For those readers who are more technically focused, this history section can be skipped without any effect on the technical development (but with a great loss of context).

The book unfolds in a clear progression. The first chapter tells the story of how the optical plate diagram came to be. It revisits Petzval's pioneering insights, Aldis's forgotten derivations, and Burch's transformative diagrams. More than a historical curiosity, this story shows how ideas evolve, fade, and return, and how design breakthroughs are often waiting to be rediscovered.

Chapter 2 teaches the basics. Starting with single lenses and mirrors, it walks the reader step by step through constructing an optical plate diagram. You will see how to move from equations to a simple drawing that reveals the structure of aberrations. Along the way, the Schmidt telescope provides a natural example of how a clever arrangement of elements can achieve dramatic correction.

With chapter 3, the book scales up to multi-element telescope systems. Here, two-, three-, and four-mirror designs are laid out, showing how optical plate diagrams capture the subtle balances that make anastigmatic systems possible. What once took decades of incremental innovation can be understood and even anticipated in a single diagram.

Chapter 4 shifts from telescopes at infinity to finite conjugate systems, the kind used in relays and imaging instruments. From the Offner relay to James Clerk Maxwell's theoretical 'perfect instrument', the diagrams reveal overlooked solutions and design opportunities. The applicability of design insights gained from the optical plate diagram is demonstrated here in the uptake of design innovations by major projects in modern astronomical optics.

In chapter 5, the method is applied to surveys of entire families of systems. It is shown how surveys can be set up to cover the entire parameter space for classes of systems, giving truly global solutions. Two-, three-, and four-mirror configurations are compared and classified by way of example. With the optical plate diagram as a guide, the reader can rapidly understand relationships that would otherwise demand extensive computational analysis.

References

[1] Lemaître G R 2009 *Astronomical Optics and Elasticity Theory: Active Optics Methods Astronomy and Astrophysics Library* (Berlin: Springer)

[2] Pretsch P 1857 On Prof. Petzval's researches in optics *J. Photogr. Soc.* **102** 107

[3] Science Museum Group Collection 2025 Brunsviga calculating machine https://collection.sciencemuseumgroup.org.uk/objects/co59679/brunsviga-calculating-machine-pinwheel-calculating-machine (accessed 10 August 2025)

[4] Wilson R N 2007 *Reflecting Telescope Optics I: Basic Design Theory and its Historical Development* 2nd corr edn (Berlin: Springer)

[5] Burch C R 1942 On the optical see-saw diagram *Mon. Not. R. Astron. Soc.* **102** 159–65

[6] Burch C R 1943 On aspheric anastigmatic systems *Proc. Phys. Soc.* **55** 433–44

[7] Thompson K P 2007 The earliest history of optical design on large computers *Proc. SPIE* **6668** 66680C

[8] Levenberg K 1944 A method for the solution of certain non-linear problems in least squares *Q. Appl. Math.* **2** 164–8

[9] Marquardt D W 1963 An algorithm for least-squares estimation of nonlinear parameters *SIAM J. Appl. Math.* **11** 431–41

[10] Meiron J 1965 Damped least-squares method for automatic lens design *J. Opt. Soc. Am.* **55** 1105–9

Chapter 1

The origins of plate diagram theory

This chapter details the background of the development of the optical plate diagram (OPD) and its proponents up until the present day. It begins with the earliest days of analytical optical design and promotes Joseph Petzval as a pioneer of a much more significant aberration theory than simply the curvature term named after him.

This story is not necessary if a purely technical understanding of the OPD is all the reader is interested in; this entire chapter can be skipped with no loss of technical coherence. For readers of the author's persuasion who like to understand the underlying connectivity of things, this chapter will be of interest.

On the other hand, for the student new to optics, please do not be put off by the various technical terms that crop up without explanation in this chapter. After this historical introduction, every term and every equation will be explained, step by step.

1.1 Joseph Petzval (1807–1891)

In 2007, the author, together with Dr Raymond Wilson, conducted a comprehensive investigation into the historical evidence supporting Joseph Petzval's priority in developing aberration theory, published in the Proceedings of SPIE in 2007 [1]. Unless otherwise indicated, specific historical references below can be found in reference [1]. The historical material presented in this section draws extensively upon that collaborative research, though it is presented here specifically within the context of understanding the development of plate diagram theory. While Petzval's pioneering work in aberration coefficients has broader significance for optical design, his methods bear particular relevance to the plate diagram approach that forms the central subject of this book.

Joseph Maximilian Petzval (known as Petzvál József Miksa in Hungarian and Jozef Maximilián Petzval in Slovak) had already established himself as a distinguished applied physicist and engineer before assuming the chair of mathematics at the University of Vienna in 1837. When artist-inventor Louis Daguerre first

doi:10.1088/978-0-7503-3099-2ch1

presented his photographic discovery, later termed 'Daguerreotypy,' to the Paris Academy of Sciences in February 1839, Petzval's colleague, the renowned Austrian physicist Andreas Freiherr von Ettingshausen, was present. Ettingshausen subsequently approached Daguerre directly to learn the process details.

Daguerre's original lens, manufactured by prominent French optician Charles Chevalier, was a slow achromatic doublet operating at approximately f/15, making it unsuitable for portrait work. In addition to its uncorrected field aberrations, the slow aperture required subjects to remain motionless for extended exposure periods. Recognising these limitations in the Chevalier lens and similar designs from manufacturers such as Plössl, Ettingshausen returned to Vienna and convinced his colleague Petzval to examine the challenges of photographic lens construction.

Despite having no documented experience in lens design or practical optics, Petzval reportedly embraced Ettingshausen's proposal with enthusiasm. He secured support from Archduke Ludwig, who provided 'Corporals Löschner, Haim, and eight skilled computing gunners' for assistance. Within six months, Petzval had created two lens designs, both featuring the same f/5 cemented achromatic front element paired with different air-spaced rear doublets.

The first design was his celebrated portrait lens, operating at f/3.6 with an aperture sixteen times larger than Chevalier's design, exceptionally fast for that era. The second was his lesser-known 'Orthoskop' at f/8.7, which covered a substantially wider field than the portrait lens and claimed distortion-free performance.

Petzval brought his designs to Voigtländer, a small Viennese optical firm, arranging for the prototype manufacture of the portrait lens. He shared detailed specifications and glass selections without contractual safeguards. This was a decision he later regretted. The lens was completed in May 1840 and tested by Anton Martin, who was developing expertise in the Daguerre process at the Vienna Polytechnic Institute under Ettingshausen's guidance. The results were reportedly excellent.

In 1841, Voigtländer entered the lens in a competition organised by the French Société d'Encouragement for photographic lens improvements. Simultaneously, Chevalier had developed his own large-aperture design called the 'Photographe à Verres Combinés à Foyer Variable'. Chevalier's empirical approach involved experimenting with various 'stock lens' combinations, achieving f/6 operation for portraits and slower speeds with wider fields using additional negative elements and stops. However, the image quality reflected the limitations of empirical design methods without fundamental optical understanding.

The Société d'Encouragement awarded Chevalier the Platinum medal over Voigtländer's Petzval lens entry, which received Silver. As Eder noted, subsequent developments proved this judgement incorrect, with the Petzval portrait lens gaining universal recognition among photographers by 1842. Despite Chevalier's higher prize, the worldwide success of the Petzval–Voigtländer lens demonstrated his design's inability to compete.

Voigtländer's commercial success with Petzval lenses was not shared with their designer, who had provided specifications without contractual protection beyond patronage from Austrian authorities. As sales flourished in the early 1840s, Petzval grew increasingly frustrated with his exclusion from the financial benefits (Voigtländer had reportedly paid him a single sum of 2000 florins, approximately $1000, in 1840). After bitter legal disputes, the partnership ended in 1845 with a complete communication breakdown. Voigtländer relocated operations to Brunswick in Lower Saxony, beyond the reach of Petzval's Austrian protectors.

By 1862, the Brunswick facility had produced 10 000 Petzval lenses, generating millions in revenue from this design alone. The introduction of the 'wet collodion process' in the early 1850s created demand for large-format wide-field lenses for landscape and architectural photography, prompting Petzval to revive his distortion-free 'orthoscope' design in 1854.

Petzval partnered with Viennese optician Dietzler to manufacture these lenses alongside portrait designs. Under Petzval's direct oversight, Dietzler produced what were considered the finest Petzval portrait lenses ever made. In 1856, Petzval and Dietzler released the 'Photographischer Dialyt', which achieved immediate commercial success. However, Voigtländer soon recognised this as Petzval's second 1840 design and began marketing their own version as the 'Orthoskope'. Voigtländer's international market dominance eventually forced Petzval to adopt the 'Orthoskope' name as well.

While the Dietzler/Petzval venture initially succeeded and earned a reputation for quality, Dietzler closed by 1862. The ongoing Voigtländer dispute and Dietzler's failure severely depressed Petzval. Additionally, in 1859, burglars destroyed his three-volume optics manuscript, representing nearly two decades of geometrical optics work, during a break-in at his summer residence. Following this loss, Petzval abandoned optics entirely, turning instead to acoustics research.

Eder described Petzval's final years as increasingly isolated and misanthropic, embittered by the Voigtländer conflict, Dietzler's failure, lack of recognition for his optical contributions, and conflicts with colleagues. He received few visitors and died from age-related infirmity on September 17, 1891.

1.1.1 Analysis of the portrait objective

Conrady provides lens specifications for the original Petzval design from von Rohr, summarised in table 1.1 [2].

Table 1.1. Petzval lens parameters from [2].

Element	Radius 1	Thickness	Radius 2	Nd	Vd	Nf-Nc
Lens el. 1	52.9	5.8	−41.4	1.517	60.0	0.008 61
Lens el. 2	−41.4	1.5	436.2	1.575	41.5	0.013 85
Air space			46.6			
Lens el. 3	104.8	2.2	36.8	1.575	41.5	0.013 85
Air space			0.7			
Lens el. 4	45.5	3.6	−149.5	1.517	60.0	0.008 61

While Conrady correctly identifies the 100 mm focal length, his stated f/3 ratio would create a front element with a negative edge thickness. His f/3.5 example produces reasonable edge thicknesses and aligns with sources indicating Petzval's original lens operated at f/3.5 or f/3.6. A layout diagram of a design drawn from this table is shown in figure 1.1. Balanced spherical aberration is shown in figure 1.2 and balanced astigmatism in figure 1.3.

Given that clear statements were made by Petzval and his colleagues that he did not use ray tracing to arrive at his designs, the existence of a well-developed aberration theory in 1839 is the only credible explanation for the correction shown here. Note that the ray-tracing methods of his day were also incapable of designing such a well-balanced optical system.

Examination reveals the portrait lens achieved remarkable performance for its time, with well-balanced third- and fifth-order spherical aberrations. The balanced orders of spherical aberration and flat tangential focal surface suggest a sophisticated understanding, not blind luck. The Petzval lens was unprecedented in its day. Creating the world's first wide-field, fast-aperture lens from scratch in six months represents an extraordinary achievement.

Substantial evidence indicates Petzval developed high-order aberration theory expressed in lens construction parameters, using it to design both portrait and landscape lenses.

Born and Wolf noted in their geometrical aberration theory chapter: 'J. Petzval, a Hungarian mathematician, successfully addressed the problem of extending Gaussian formulae with higher-order angle and inclination terms. Unfortunately,

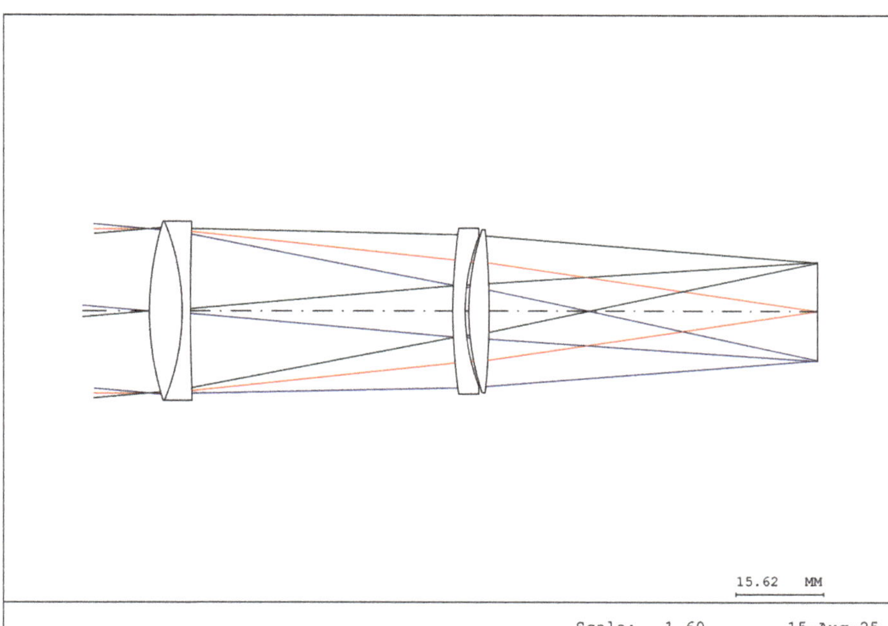

Figure 1.1. Petzval lens produced from the data in table 1.1 using Code V™ software. The entrance pupil (with a stop on the front surface) is 29 mm in diameter (f/3.4). Created with Code V.

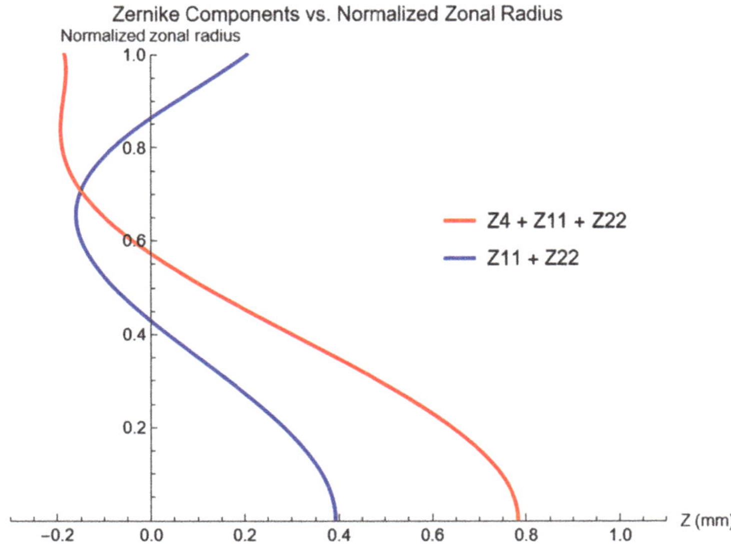

Figure 1.2. Wavefront spherical aberration in the 3rd and 5th orders, obtained by fitting high-order Zernike terms to the wavefront at the exit pupil, shows near-perfect balance in the Petzval lens with the parameters given in table 1.1. Created with Code V.

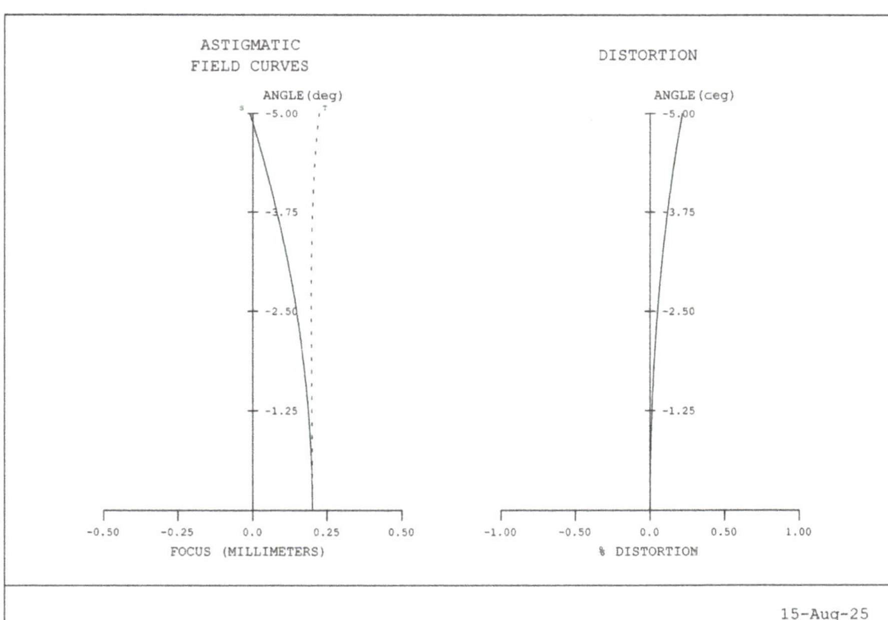

Figure 1.3. The system has a flat tangential focal surface, with the subtle curve indicating some third- and fifth-order astigmatism balancing is also at work. Created with Code V.

thieves destroyed Petzval's extensive manuscript; knowledge of this work comes mainly from semi-popular reports.'

The only plausible primary source of the 'semi-popular' reports Born and Wolf allude to, without reference, is Petzval's own books, the first published in 1843 [3] and the second in 1857 [4], where he explicitly mentioned the development of a multi-order aberration theory and provided correct aberration counts for the third, fifth, and seventh orders (with 5, 12, and 20 distinct aberration types for these orders, respectively). Unfortunately, with the notable exception of formulae for the curvature of images that now bear his name, these books were descriptive rather than prescriptive; Petzval intended to publish his formulations in more detail but, as mentioned previously, that work was lost or destroyed.

If this thesis, given in more detail in reference [1], proves correct, Petzval deserves credit for discovering not only 'Gaussian optics' and 'Seidel aberrations' but also for developing higher-order aberration theory 60 years before the first published results. Definitive proof would require discovering previously unknown papers, letters, or class notes containing Petzval's complete theoretical descriptions. On a recent visit to Vienna, the author began enquiries in that direction.

While Petzval's aberration balancing was not perfectly executed by modern standards, the fact that he achieved a result at all is the remarkable point. A decade after our 2007 paper (reference [1]), another paper was published providing a further description of Petzval's achievement [5].

Petzval's 'lost aberration theory' has a direct bearing on the topic of this book. As we shall see next, Aldis [6] apparently had direct access to this lost work and reproduced it in a paper discussed in the following section. Later too, we report an offhand comment by Schwarzschild, as well as a modern analysis that provides another strong corroborative piece of evidence (see section 3.6 of this book).

1.2 Lancelot Hugh Aldis (1870–1945)

Given his prominence in early 20th-century British optics circles, there is a surprising paucity of readily available information regarding H L Aldis. The biographical information given below comes from one primary source [7], unless indicated otherwise in the text. All of the technical information in this section, including reproduced text and diagrams, comes from reference [6].

Aldis (who went by the name 'Hugh') was the first of nine children, born in Calcutta, India. His father and two uncles had graduated from Trinity College, Cambridge, and were respectively sixth, senior, and second Wranglers.[1] In fact, one of his uncles, William Steadman Aldis, went on to become a well-known professor of mathematics and astronomy in Auckland, New Zealand, and was influential in a significant cultural change that allowed women and Maoris to obtain tertiary degrees for the first time in New Zealand [8]. So, it is not surprising that Aldis himself studied for the mathematical tripos at Cambridge.

[1] Wrangler: Title historically used at the University of Cambridge for a student attaining first-class honours in the Mathematical Tripos; the Senior Wrangler was the highest-ranked candidate.

During his time at Cambridge, Aldis developed an interest in astronomy and made himself a 6′ Newtonian telescope, grinding, polishing, and figuring the primary mirror himself by hand. It seems that his enthusiasm for amateur telescope-making distracted him from his studies, and his parents were disappointed when he graduated as 14th wrangler. He himself, however, had expressed contempt for the way geometrical optics had been taught at Cambridge, and his subsequent career proved his point; within a few years, while in the employment of Messrs. J.H. Dallmeyer Ltd. and using his own methods of computation, he produced a series of anastigmatic lenses known collectively as the 'Dallmeyer Stigmatic' lenses, which gave good corrections over a field of view of 60°. Series III of this line comprised two singlet front elements and a cemented doublet rear element, providing the bulk of the optical power. This design was patented, completely anticipating the Zeiss Tessar, developed in parallel by Paul Rudolf and subsequently patented in 1902.

In 1912, in a deliberate provocation of Zeiss, he released a version of his Series III design with an f-number that exactly matched that of the patented Tessar. Following the success of this lens, Zeiss became aware of it and sued for patent infringement, but proceedings collapsed due to the 'manifest anticipation (of the Tessar) by the Series III lens'. By this time, Aldis had formed an optical company with his brother, A. C. W. Aldis, which had much success, developing signalling lights for the Royal Navy that are used to this day (Aldis lights) and an innovative and successful series of rifle scopes and gunsights.

In 1900, Aldis wrote a profound paper that was published in The Photographic Journal. 'On the Construction of Photographic Objectives' (see reference [6]) provided a full derivation of third-order aberrations obtained solely by considering rays in the caustic curve. Remarkably, Aldis attributed the complete derivation to 'Petzval, 1840' and was obviously referencing, in detail, a publication of Petzval's that is now lost. A diagram apparently reproduced by Aldis from a lost work of Petzval's is shown in figure 1.4.

Two key paragraphs from this derivation are reproduced below to convey the idea. Interested readers are encouraged to consider the referenced paper, bearing in mind that the full derivation of 'Seidel aberrations' is directly attributed to a lost work of 'Petzval, 1840', which predates Seidel by 16–18 years. Referring to figure 1.4, we have:

> Consider first rays lying in the plane AOP which we call the primary plane. After refraction these rays will still remain in the primary plane, but these rays after refraction will no longer pass through a point, but will touch a caustic curve $Qq_2q_3q_4$ the marginal rays of the pencil such as PR_2 coming to a shorter (or longer as the case may be) focus Q_4R_4, than the central rays of the pencil PR_1. This well-known effect is described as spherical aberration and as a measure of the spherical aberration, we take the rate at which the focus Q_1 recedes from the geometrical focus Q as the angle RPR_4 increases.
>
> We next come to the consideration of what we call 'coma'. In the previous figure, we supposed that rays of light were proceeding in all directions from the point P, but in practice the rays of pencils which go to

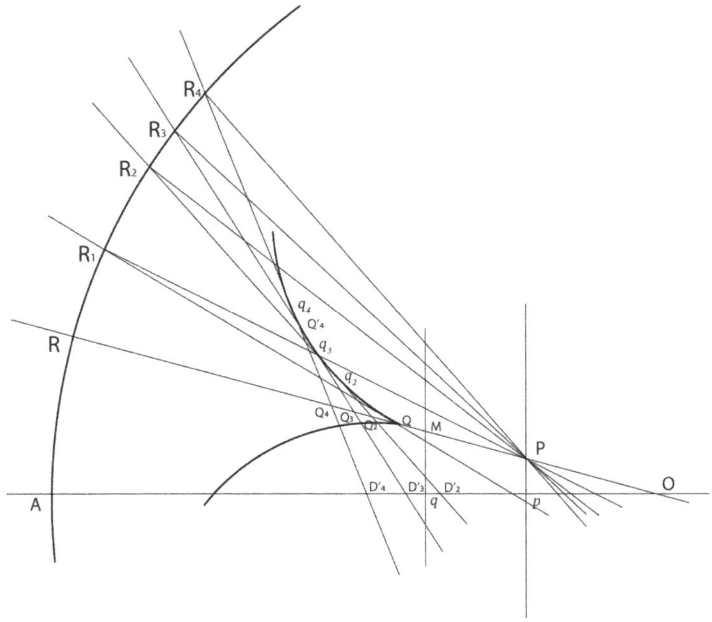

Figure 1.4. Figure one from Aldis's paper, in which he walks through the complete derivation of third-order aberrations from the caustic curve and attributes this derivation in its entirety to 'Petzval, 1840'. Such a derivation has remained unknown to the field of optical aberration theory until the modern day.

make up the image must be limited by a stop of some sort which is a small round aperture the centre of which is on the axis of the system. Now in this figure the spherical aberration is symmetrical about the line OP passing through the centre of curvature of the surface. If then we limit the incident pencil by means of a stop whose centre is at O_2 the spherical aberration after refraction will be of the regular pattern symmetrical about the central ray of the pencil. If however we limit the pencil by means of a stop at D_3 away from the centre of the surface the emergent pencil will consist of the bundle of rays between q_2R_2 and q_4R_4 which touch the caustic ray away from the vertex, and moreover the spherical aberration is not symmetrical about the central ray R_3q_3. This unsymmetrical spherical aberration of an obliquely incident ray is called coma and the amount of coma evidently depends on the obliquity of the angle of incidence of the axis of the pencil of incident rays (the angle R_3PR). In fact **Petzval** shows that a measure of the coma is the quantity

$$L\xi$$

where L is the measure of spherical aberration for a centrally incident pencil and ξ is a quantity depending on the distance of the centre of the stop D_3 from O.

It is worth noting a few points here. This derivation is not complete (Aldis refers to missing detail '...Petzval shows that...') but is enough to point exactly to the complete derivation. Also, Aldis's figure 1.1 (figure 1.4) has some mistakes; for example, the points D_2–D_4 are not shown (but can be inferred), and the images D'_2–D'_4 are incorrectly shown to lie on the axis of the refracting surface, whereas they are meant to be images of the edges of the stop at D.

The derivation continues for several pages, and Aldis concludes his derivation for a single surface, referencing his figure 1.3 (figure 1.5). The text is long out of copyright and is of significant historical interest, and the paper is difficult to locate today, so Aldis's conclusions are presented here verbatim:

> Hence, we see that in the case of oblique refraction at a single spherical surface, the pencil after refraction has:
> Spherical aberration measured by L
> Coma measured by $L\theta$
> And the image has primary curvature $-U-3L\theta^2$
> secondary curvature $-U-L\theta^2$
> and distortion $\frac{1}{2}(U + L\,\theta^2)\,\theta$
> Since the primary curvature is $U + 3L\theta^2$
> and secondary curvature is $U + 3L\theta$

We may take it that U is a measure of the curvature and $L\theta^2$ of the astigmatism. With respect to the quantity U we may remark that if the object lies in a sphere centre O, the image must also lie in a sphere centre O, and hence in this case $\frac{1}{OP}$, $\frac{1}{OQ}$ will be the curvature of the object and image respectively. By the law of geometrical foci

$$\frac{1}{\mu' OQ} - \frac{1}{\mu OP} = \left(\frac{1}{\mu'} - \frac{1}{\mu}\right)\frac{1}{r}.$$

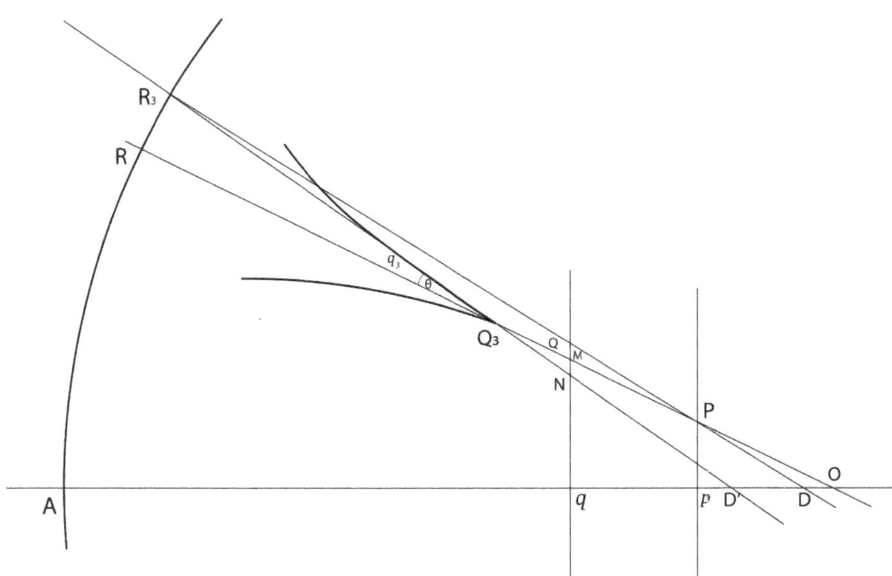

Figure 1.5. Quantities referred to at the conclusion of Aldis's description of Petzval's derivation.

Hence in this case

$$\frac{\text{curvature of image}}{\mu'} - \frac{\text{curvature of object}}{\mu} = -\frac{\mu' - \mu}{\mu'\mu}p.$$

This relation will always hold between the curvature of the image and object for pencils through O. Hence, if the object be flat

$$\frac{\text{curvature of image}}{\mu'} = -\frac{\mu' - \mu}{\mu'\mu}p = -U \text{ aforesaid.}$$

We have seen thus how these five aberrations arise in the refraction of a pencil at a single spherical surface. Now, as **Petzval** shows if a pencil of rays from a point on the object be refracted by an optical system consisting of any number of such surfaces, the

1. Spherical aberration of the emergent pencil is measured by

$$\sum L_n$$

2. Coma of the emergent pencil is measured by

$$\sum L_n \xi_n$$

3. Primary curvature of the image is

$$-\sum U_n - 3\sum L_n \xi_n^2$$

4. Secondary curvature of the image is

$$-\sum U_n - \sum L_n \xi_n^2$$

5. Distortion of the image is

$$\sum U_n \xi_n + \sum L_n \xi_n^3$$

Where L_n is the spherical aberration of the nth surface, ξ_n is a measure of the obliquity of incidence on that surface of the axis of the pencil and $U_n = \frac{\mu_n - \mu_{n-1}}{\mu_n \mu_{n-1}} p_n$, p_n being the curvature of the nth surface. Hence, if the emergent pencil is to be perfectly free from spherical aberration, coma, and astigmatism, and if the image is to be plane and non-distorted, we must satisfy the five conditions:

1. $\sum L_n = 0$ (spherical aberration)
2. $\sum L_n \xi_n = 0$ (coma)
3. $\sum L_n \xi_n^2 = 0$ (astigmatism)
4. $\sum U_n = 0$ (curvature)
5. $\sum U_n \xi_n + \sum L_n \xi_n^3 = 0$ (distortion)

The quantity ξ_n, as we have said, measures the obliquity of incidence on the nth surface of the axis of the pencil forming the image; in other words, the form of the above equations will depend on the position of the stop which we take to limit the incident pencils. But we shall find on taking the general case that if we satisfy the equations for any other assumed position. Hence, we may evaluate the quantities ξ_n for any arbitrary position of a stop we please, and it is usual to take this arbitrary position to be at the vertex of the first surface.

We should pause here and consider the weight of meaning in what we have just read. Here, we have one of the most talented optical designers in pre-war 20th-century Britain and arguably in the world at that time. Aldis's anastigmat design anticipated Rudolph's famous Tessar design [9], and history shows that a fiercely protective corporation, Zeiss, dropped its patent case against Aldis, validating his priority (see reference [6]). Aldis was proud and sure enough of himself to put himself in legal jeopardy with Zeiss in the first place by deliberately releasing a lens in exactly the same format as the Zeiss lens and then not backing down in response to initial cease-and-desist communications, provoking a court case. And here, in the paper referred to above, Aldis directly attributes the complete and detailed derivation of his aberration theory, which he reproduces in detail, to Joseph Petzval, 1840. Such a man who would challenge Zeiss would hardly fail to claim this innovative derivation for himself if he had produced it. We have no reason to doubt his words.

The derivation Aldis presented was based explicitly on the geometric properties of the caustic. The earliest known work relating Seidel aberrations to caustics is from 1990: 'Ray trajectories and caustic: Clairaut's equation' [10] by Ojeda-Castañeda and López-Olazagasti, which used known Seidel coefficients to generate caustic surfaces. The first published derivation of a Seidel coefficient *from* caustic geometry appears in 2010, in the paper 'First full analytical derivation of Seidel spherical aberration from caustic' [11], but this was limited to spherical aberration alone and did not include off-axis terms or stop-shift effects.

In contrast, Aldis's paper presents a complete derivation of all third-order aberrations obtained directly from an examination of the caustic and uniquely states the dependence of these aberrations on the position of the aperture stop. This caustic derivation is a 'technological fingerprint' that does not exist in the technical literature outside Aldis's paper to this day.

In the rest of his paper, Aldis proceeds to prove that a minimum requirement for producing an anastigmatic lens is four optical surfaces and uses the formulation given above to produce an example of a two-element monochromatic anastigmat. The success of Aldis's application of Petzval's theory to lens design is confirmed by the output of his company, such as highly successful camera lenses, gunsights, the Aldis signalling lamp, etc. Just as with Petzval, the application of an aberration-theory-based approach to optical design gave Aldis Brothers Limited a strong competitive advantage in a world where most optical designers did not have access to this approach.

It is a matter of historical curiosity, given the strongly interconnected nature of British optics at the start of the 20th century, and also strong international links with the German, US, French, and Austrian research in the field, that more was not made of Aldis's work. This paper was not cited by Conrady, T. Smith, von Rohr, or any of the others who should have realised its significance immediately. In fact, the first, and prior to the author [12, 13], the only person to discover and reference Aldis's aberration theory work in a publication was Cecil Reginald Burch, who is discussed in the following section.

1.3 Cecil Reginald Burch (1901–1983)

In this section, all biographical information and any technical material not otherwise cited are drawn from a single source: a 39-page biographical memoir published on the occasion of Burch's death by his friend, Thomas Edward Allibone [14]. For anyone wanting to know more about the remarkable career and character of Burch than can be found below, this reference is recommended reading.

Cecil Reginald Burch was the archetype of a polymath. Over his career, he made significant innovations in the various fields of physics in which he found himself active. His career spanned an 11-year stint at Metropolitan Vickers, two years at the Physics Department at Imperial College London as a Leverhulme Fellow in Optics, and the rest of his life at Bristol University, first as a Research Associate, then as a Fellow of the H H Wills Physics Laboratory, and from 1948 to 1966 as a Royal Society Warren Research Fellow. He continued to do research in the laboratory up until the last few weeks of his life.

While at Metropolitan Vickers, he invented, together with Neville Ryland Davis, the electric induction furnace, which remains the main workhorse of metal foundries the world over to this day [15].

While doing associated work at high vacuum, Burch had realised that he could distil oils that could replace mercury in rotary pumps, essentially inventing the oil diffusion pump, which led to several patents and a paper in Nature [16]. His work in vacuum found immediate widespread industrial application, and he maintained a supervisory role with the company commercialising the technology, F. E. Bancroft. While there, almost in passing, he pioneered the newly discovered process of aluminising optical surfaces *in vacuo* [17], which was later to become a mainstay for astronomical telescope mirrors. Every large astronomical telescope mirror in the world today working in the optical and near-infrared regions relies on vacuum-deposited metal coatings, for which the necessary high vacuum is achieved using oil diffusion pumps first developed by Burch.

His work on high-power radio valves, whose production relied on the higher-quality vacuum he had developed, produced outstanding results. These valves laid the basis for 'Chain Home', the British aircraft detection radar system that uniquely contributed to victory during the Battle of Britain. Regarding Chain Home, the eminent historian Zimmerman says, '*it was already a very close-run affair, and the whole course of World War 2 could have been different*' [18].

Further to Burch's contributions to the war effort, we have the following from Allibone:

> …his diffusion pump operating with oil had made it possible to build the large valves, and hundreds of extremely large oil-diffusion pumps were in use to evacuate the vessels in which beams of 235U and 238U ions pursued their separate courses in a magnetic field in the uranium separation plants at Oakridge in 1944/45. Ernest Lawrence, who was responsible for this huge venture, could not speak too highly of the value of Burch's contribution to the war effort.

In 1933, Burch's brother, Francis, with whom he had lived and worked throughout his early career, died suddenly. This tragedy led Burch to make a course change in life, and he applied for and was granted a Leverhulme Research Studentship at Imperial College under Professor G. P. Thomson. He left Metropolitan Vickers Co. in September of that year, being admitted to conduct research 'On the production of aspherical optical surfaces and on their imaging properties in combination'. Soon after beginning at Imperial College, a colleague from Shell, who had collaborated on the mass production of vacuum oils developed by Burch, informed him of the work by Zernike at Groningen, who had at that time developed his phase contrast method for microscopy but had not yet published. Burch immediately realised that the same principle could be applied to the testing of optical foci and, in a moment, had invented the Zernike test, which proved to be a remarkable improvement on the Foucault test, the main means of testing aspherics at that time. Burch wrote to Zernike saying:

> 'van Dijk told me of your work and this is what I have done, may I publish it?' He replied 'By all means publish it and since I have published nothing on the method I will write a paper on the theory, and both papers can come out together'.

Within days, Burch had devised means to make Zernike phase-retarding discs 0.01 microns in diameter and immediately applied the technique, quickly demonstrating the capability to measure figures of aspheric surfaces to an accuracy of 0.1 fringes [19, 20].

In Allibone's words:

> He sent the substance of his results quickly to Zernike asking his permission to publish, and it was published by the Royal Astronomical Society immediately after Zernike's paper. Previously the knife-edge test had been standard; now Burch proposed the Z -test as a null test which had so successfully shown up errors of 0.1 of a fringe.

Zernike, who went on to win the Nobel Prize in Physics (1953) for his invention of the phase contrast method, sent one of his early microscopes to Burch. Burch immediately turned his attention to developing his own reflecting microscope objectives and rapidly became the world leader in this area.

Remarkably, the 'optical see-saw diagram', otherwise known as the 'optical plate diagram' and the subject of this book, barely gets a mention in reference [13]; it is discussed twice, first on page 25, with respect to an optics patent he had filed, and a related paper 'On the optical see-saw diagram' [21], and second on page 36, in the context of Burch's last paper on the subject, which he presented at the age of 78 at an optics conference in Bath organised by his son, Dr J M Burch [22]. The introduction to this paper bears repeating here:

> When Dr. J. M. Burch asked me to give this talk, I protested, 'The Plate Diagram is old work-36 years old. Conferences ought to be about new work!' He said, 'But they ought to know about it, and they don't.' I retorted, 'Then they can bloody well go and read my published papers!' He said, 'But they wouldn't know where to find them; and anyway, they would have no time to read them!' Well, he has a point there-if we all read all the papers we ought to, we wouldn't get any work done at all. So, I consented.

This extract serves to illustrate a little of the remarkable character of this giant in 20th century optics.

Burch had first started thinking about what was to become the plate diagram on learning of the Schmidt telescope, invented by Bernhard Schmidt in 1930 [23]. Already, in 1938, he had used an aspheric plate he had made himself to improve the performance of an existing lens [24]. In 1943, he described how this experience led him to think in detail about how Seidel aberrations arise at each surface of an optical system [25]. As we shall see in chapter 2, the Schmidt plate embodies the essence of the transform that Burch pioneered.

This material will be covered carefully, with all necessary derivations and diagrams in the next chapter. It should be noted, however, that the method developed (independently) by Burch **exactly** reproduces the results reported by Aldis and attributed to Petzval, albeit arrived at via a different path. While the former method reached its conclusions via direct consideration of the caustic, in this case, Burch introduces the elegant concept of an 'astigmatising plate' that perfectly cancels the anastigmatising Schmidt plate (the 'lamina of retardation proportional to the 4th power' of the zonal radius).

In fact, soon after this paper was published, Burch produced a second paper on the topic, 'On aspheric anastigmatic systems' [25], in which he reports his discovery of Aldis's paper:

> I have since found that the basic underlying idea is by no means new, and that the see-saw diagram for a system of spherical surfaces was given by H. L. Aldis no less than 43 years ago. 'Science moves but slowly, slowly creeping on from point to point'. I shall now describe the method in its generalized form de novo....

Here, we see Burch both graciously acknowledging Aldis's priority and understating his own contributions. In this second paper, Burch immediately grasps Aldis's theorem regarding the minimum number of non-concentric surfaces required to produce an anastigmatic correction and presents it in a somewhat simpler form. This will be described in detail in chapter 3.

1.4 To the present day

In his heyday, Burch propagated his method directly to numerous students and colleagues. Most notably, Dr Edward Linfoot, a colleague of Burch's at Bristol (who moved into a career in optics following advice from his psychiatrist that he should try to find something more 'down-to-earth' to do than pure mathematics), produced the seminal work 'Recent Advances in Optics' [26]. To this day, this book remains the only one to describe the OPD in detail. In it, Linfoot uses the plate diagram to demonstrate or derive numerous systems, including the Schmidt–Cassegrain designs that have become popular in amateur astronomy. This book also reports for the first time the complete diffraction theory of the Foucault test, which had been derived by one of Burch's students, Ben Gascoigne.

In fact, one of the main wavefront sensors used in modern adaptive optics systems is the pyramid wavefront sensor (PWS), first proposed by Roberto Ragazzoni [27]. When the author met one of Ragazzoni's colleagues, Dr Armando Riccardi, who was at that time busy installing an adaptive optics system on the Large Binocular Telescope [28], Riccardi mentioned that taking diffraction into account with the PWS was critical. The author inquired of Riccardi whether he knew of 'Recent Advances in Optics', as it contained exactly this analysis, to which Riccardi replied, '*that was my Bible!!!*'.

The only other significant new design to be produced using this method prior to 2000 was the 'Gascoigne plate'. In a brief note to The Observatory in 1965 [29], Gascoigne describes his idea of essentially 'reimaging' a Schmidt-like plate to near the focus of a 2.5 m Ritchey–Chrétien telescope to correct the astigmatism that otherwise limited the field of view to a radius of about 25 arcminutes.

Gascoigne's plate, near the focal plane, images to a long distance in the object space of a telescope. Considering that astigmatism grows as the square of the plate distance from the pupil, whereas coma is linear and spherical aberration has no x-dependence (see the equations in the Aldis section), we see that a plate with a very weak spherical aberration contribution, positioned at a large value of x, could provide useful amounts of astigmatism correction while only mildly affecting the correction of spherical aberration or coma.

Gascoigne also pointed out that if the telescope itself was designed with such a small aspheric plate in mind at the outset, the asphericities on the primary and secondary mirrors could be adjusted together with the plate to produce an anastigmat.

The world's first modern wavefront sensor, the Shack–Hartmann wavefront sensor, has a connection to Gascoigne's work. In the late 1960s, Aden Meinel first proposed to his colleague, Roland Shack, that an interesting optical test would be possible if one reimaged a Hartmann mask to a small size [30] (reference [30] also applies to the following paragraph).

It is likely that Aden Meinel was himself influenced by the work of Ben Gascoigne, whose idea of applying Schmidt-like plates close to the focal plane to correct Ritchey–Chrétien astigmatism had recently been applied to the No. 1 36′ telescope at the nearby Kitt Peak National Observatory. Gascoigne and Meinel had collaborated on numerous projects; for example, in the site testing that led to the selection of Kitt Peak as the site for the National Observatory, and Meinel was director of the observatory at that time. The 'relayed pupil Hartmann Test' concept is clearly closely related to the 'relayed Schmidt plate corrector'.

Subsequent to Gascoigne's brief note in 1965, the first instance the author is aware of in the literature is Simon [31], who only mentions that it was used without giving any detail, then Martha Rosete–Aguilar [32], who gives a more detailed discussion of two-mirror telescopes citing Burch and Gascoigne.

Surprisingly, this remarkable method saw little adoption. In Aldis's time, it proved its worth, as his company produced significantly superior optical designs to those of competitors. Later, Burch, together with Linfoot and Gascoigne, clearly demonstrated the power of the method as a superior visualisation tool and used it to develop significant optical designs for reflecting microscope objectives, telescopes, and correctors. Burch himself expressed some bitterness about this in his later years.

A possible reason for the lack of uptake is the source material. The Aldis paper is in an obscure journal from 1900, which no one today would find unless they looked for it explicitly, and even then, it is not easy to locate. Burch's papers are now also very dated. In both cases, while the overall idea is clearly communicated, a lot of detail is skipped, making it difficult for people not already experts in ray tracing and aberration coefficient calculation to quickly pick up and use.

This book aims to rectify that situation by providing a detailed, step-by-step guide to the calculation and application of plate diagrams in optical design and analysis, so that more people may share in the benefits the author has experienced.

References

[1] Rakich A and Wilson R N 2007 Evidence supporting the primacy of Joseph Petzval in the discovery of aberration coefficients and their application to lens design *Proc. of SPIE 6668: Novel Optical Systems Design and Optimization X* 66680B (SPIE)

[2] Conrady A E 1960 *Applied Optics and Optical Design, Part 2* (New York: Dover Publications) 805–10

[3] Petzval J 1843 *Bericht über die Ergebnisse einiger dioptrischer Untersuchungen* (Pest: Beim Verlag von C.A. Hartleben)

[4] Petzval J 1857 *Bericht über optische Untersuchungen Denkschriften der Kaiserlichen Akademie der Wissenschaften, Mathematisch-Naturwissenschaftliche Classe* vol 13 (Wien: Kaiserlich-Königliche Hofund Staatsdruckerei) 197–220

[5] Sasián J 2017 Joseph Petzval lens design approach *Proc. SPIE–OSA 10590 Int. Optical Design Conf.*

[6] Aldis H L 1900 On the construction of photographic objectives *Photogr. J.* **30** 291–9

[7] Aldis A C W 1946 Obituary notices *Mon. Not. R. Astron. Soc.* **106** 26–7

[8] Tee G J 1998 Professor and Mrs. Aldis: mathematics, feminism and astronomy in Victorian Auckland *South. Stars.* **38** 18–27
[9] Rudolph P 1903 Photographic objective *US Patent* 721240
[10] Ojeda Castañeda J and López Olazagasti E 1990 Ray trajectories and caustic: Clairaut's equation *Microwave Opt. Technol. Lett.* **3** 375–8
[11] Avendaño Alejo M, González Utrera D and Castañeda. L 2010 Properties of caustics produced by a positive lens: meridional rays *J. Opt. Soc. Am.* A **27** 2252–60
[12] Rakich A 2001 A complete survey of three mirror anastigmatic reflecting telescope systems with one aspheric surface *M.Sc. Thesis* (University of Canterbury, Christchurch, New Zealand)
[13] Rakich A and Rumsey N 2002 Method for deriving the complete solution set for three mirror anastigmatic telescopes with two spherical mirrors *J. Opt. Soc. Am.* A **19** 1398–405
[14] Allibone T E 1984 Cecil reginald burch, 12 May 1901—19 July 1983 *Biogr. Fell. R. Soc.* **30** 3–42
[15] Davis N R and Burch C R 1925 Circuit for induction heating furnaces *British Patent* No. 257021
[16] Burch C R 1928 Oils, greases, and high vacua *Nature* **122** 729
[17] Strong J 1934 Evaporated aluminum films for astronomical mirrors *Publ. Astron. Soc. Pac.* **46** 18–27
[18] Zimmerman D 2010 *Britain's Shield: Radar and the Defeat of the Luftwaffe* (Stroud: Amberley Publishing)
[19] Burch C R 1934 On the phase contrast test of F. Zernike *Mon. Not. R. Astron. Soc.* **94** 384–99
[20] Burch C R 1934 Address on Zernike's test for mirrors *J. Br. Astron. Assoc.* **44** 328
[21] Burch C R 1942 On the optical see-saw Diagram *Mon. Not. R. Astron. Soc.* **102** 159–65
[22] Burch C R 1979 Application of the Plate Diagram to reflecting telescope design *Opt. Acta* **26** 493–504
[23] Schmidt B 1931 Ein lichtstarkes komafreies Spiegelsystem *Mitt. Hamburger Sternw.* **7** 15–21
[24] Burch C R 1938 Note on improving the covering power of a photographic lens by means of an aplanatising plate *J. Br. Astron. Assoc.* **48** 213–6
[25] Burch C R 1943 On aspheric anastigmatic systems *Proc. Phys. Soc.* **55** 433–44
[26] Linfoot E H 1955 *Recent Advances in Optics* (Oxford: Clarendon)
[27] Ragazzoni R 1996 Pupil plane wavefront sensing with an oscillating prism *J. Mod. Opt.* **43** 289–93
[28] Esposito S *et al* 2011 First light adaptive optics system for LBT: FLAO#1 commissioning *Astron. Astrophys.* **528** A56
[29] Gascoigne S C B 1965 Optical systems for large telescopes *Observatory* **85** 24–7
[30] Rakich A 2020 The impact of Roland Shack's wavefront sensor on the development of modern active optics *Proc. of SPIE 11451: Advances in Optical and Mechanical Technologies for Telescopes and Instrumentation* 1145102 (SPIE)
[31] Simon J M 1993 A new high-resolution monochromator with spherical mirrors *Proc. SPIE* **1983** 174–6
[32] Rosete-Aguilar M, Martinez-Garcia M, Malacara D and Malacara-Hernandez D 2000 Optical design of two-mirror telescopes with spherical mirrors *Appl. Opt.* **39** 5818–25

IOP Publishing

Analytical Lens Design using the Optical Plate Diagram
An introduction to the fundamentals with practical applications
Andrew Rakich

Chapter 2

The plate diagram for a single surface

Following the stated intentions of this book, to allow anyone with a reasonable grasp of elementary mathematics to construct and understand a plate diagram, this chapter explains, step by step, the various calculations required to construct a plate diagram from a single spherical or aspheric (rotationally symmetrical) surface. Readers who are already familiar with this material may still be interested to see how it is presented; alternatively, they may wish to skip ahead to the section on the calculation of aberrations in Schmidt telescopes.

Concepts and derivations are given in full here and will be useful for people unfamiliar with the material, but for those impatient to go straight to producing plate diagrams, all that will be required from this chapter will be the equations referred to in the last section.

2.1 First-, third-, and high-order

For those readers who are not necessarily students or practitioners of optical design, it is first necessary to define what we mean by 'Gaussian optics', 'first-order', 'paraxial', 'extended-paraxial', 'third-order', and 'high-order', as these terms come up a lot, and several of them are directly interchangeable.

Consider figure 2.1.

As is customary, i and i' represent the angles made by the rays with respect to the surface normal at the point of contact with the optical surface before and after refraction (or reflection), respectively. Likewise, n and n' represent the refractive indices of a homogeneous optical medium before and after refraction (or reflection).

In the case of reflection in a homogeneous medium, $n' = -n$.

Snell's law states that:

$$n \sin i = n' \sin i'. \tag{2.1}$$

doi:10.1088/978-0-7503-3099-2ch2

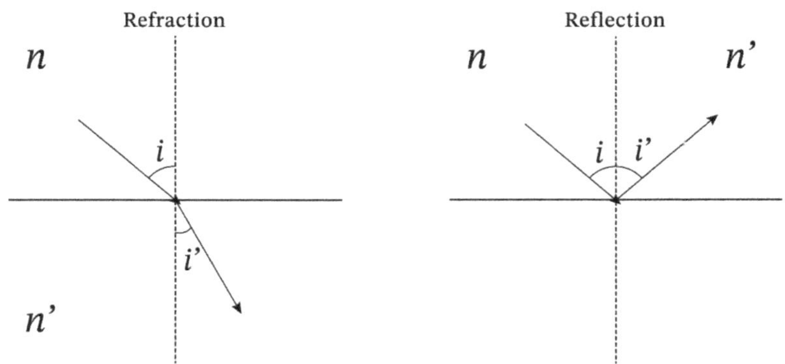

Figure 2.1. Refraction and reflection at an optical surface.

The Taylor series expansion of $\sin\theta$ about $\theta = 0$ is:

$$\sin\theta = \sum_{k=0}^{\infty}\frac{(-1)^k}{(2k+1)!}\theta^{2k+1}. \tag{2.2}$$

Expanding the sum gives:

$$\sin\theta = \theta - \frac{\theta^3}{3!} + \frac{\theta^5}{5!} - \frac{\theta^7}{7!} + \frac{\theta^9}{9!}. \tag{2.3}$$

In this expansion, we consider three divisions. In the first division, we consider only the first term of the expansion, θ, and neglect all terms with powers of $\theta>1$.

This simplification of equation (2.3) gives:

$$\sin\theta = \theta. \tag{2.4}$$

Snell's law from equation (2.1) becomes:

$$ni = n'\,i', \tag{2.5}$$

and this is completely valid for infinitesimal angles. By similar arguments we have $\cos\theta = 1$, and therefore $\tan\theta = \sin\theta$.

Karl Gauss developed a complete system of optics based on this approximation of Snell's law [1], which is now referred to as 'Gaussian optics'. Alternatively, optical relations which are only valid within this small-angle approximation are referred to as 'first-order', for reasons that should now be obvious. Considering an optical system in which all ray angles are infinitesimal, it follows immediately that all ray heights are likewise infinitesimal and rays are only infinitesimally differentiated from the optical axis. From this, we get the name 'paraxial optics'.

In this sense, the terms 'Gaussian optics', 'first-order optics', and 'paraxial optics' all mean exactly the same thing.

Next, we consider the cubic term in the expansion of $\sin\theta$: $\frac{\theta^3}{3!}$. If we allow our infinitesimal angles to increase to a range of values in which these cubic terms now have an influence on ray paths, but terms of the 5th order and higher do not, we have entered the 'extended paraxial region'. This extended paraxial region, or 'third-order' optics, was discussed in detail in a historical context in chapter 1. History records that Seidel performed the first complete analysis of third-order optics, identifying five distinct aberrations that only occur in this region [2]. For this reason, third-order aberrations are commonly referred to as 'Seidel aberrations'. As was discussed in chapter 1, there is strong circumstantial evidence that Seidel was preceded in this by Joseph Petzval. In fact, the optical derivation from the caustic curve given by Aldis in section 1.2 and attributed to Petzval goes beyond Seidel, showing completely the dependencies of these five aberrations on the system aperture stop position.

The third division of the Taylor series expansion of sin θ contains all terms with powers of θ greater than three (five, seven, nine…). Perturbations to ray paths that are dependent on these orders of the expansion are referred to as 'high-order aberrations'.

The optical plate diagram can be considered as a transform operating in the extended paraxial region, i.e. accurate to the third order. As will be shown, we can start with any optical system of lenses and/or mirrors, where both first-order and third-order quantities are calculated, and produce a plate diagram from it.

The plate diagram is composed entirely of optical elements that are flat to a first order of approximation and have departures from flatness that only affect ray paths to the third order of the angle expansion. The plate diagram does not discard the first-order properties of a real optical system; rather, it 'encodes' them in the positioning of the individual third-order plates that make up the system. In terms of analysis, the plate system only perturbs rays at the third-order level. Therefore, all of the first-order mathematics that is necessarily included in a third-order analysis of a normal optical system is removed from the plate diagram analysis. To paraphrase Burch, 'a dark miasma of algebraic fog is lifted', revealing how each element affects and interacts with the overall system.

This simplicity would seem to be lost when we extend considerations to the 5th and higher orders of expansion. In the case of the third-order aberrations, these arise simply at each surface and sum for the system. When we consider high-order aberrations, additional complexity arises. An aberrated wavefront, interacting with successive optical surfaces in the system, generates additional aberrations at each surface. Low-order aberrations create high-order 'ripples' that must be accounted for as well. Again, a 'miasma of algebraic fog' arises. Nonetheless, it is an interesting direction to consider, and future work may show the advantages of a plate diagram approach to high-order analysis, but these have not been demonstrated to date.

A final point to mention in this section concerns a difference between reflecting systems and refracting systems. The deflection of a ray parallel to the optical axis at some fixed height is proportional to the curvature of the surface and to the change in refractive index at the surface. Therefore, as the change in refractive index increases, we can achieve the same deflection with less curvature, which, in this case, also means with smaller incidence angles. This can be thought of as 'refractive efficiency'. The refractive efficiency of an optical surface increases as the change in refractive index at the surface increases. With typical optical materials, the change in refractive index at an air-to-glass surface might range between 0.45 and 0.75. But in the case of reflection, with a mirror in air, and remembering that the refractive index simply reverses sign on reflection, the change in refractive index has a magnitude of two.

The consequence of this is that a reflecting surface in air achieves the same deflection as a refracting surface in air, with between a fourth and a third of the curvature and correspondingly smaller angles of incidence. Considering what we have discussed so far about orders of aberration, it should be clear that a reflecting surface generally imparts significantly less high-order aberration to a ray than a refracting surface of the same optical power.

For this reason, the optical plate diagram has been used almost exclusively in the analysis of reflecting and/or catadioptric systems with real plates. The third-order approximation is just that much more accurate compared to the case of a refracting system with equivalent optical power.

While the plate diagram transform can usefully be applied to refracting systems, especially those of relatively low optical power, such as telescope wide-field corrector lenses, this use is not as common. Therefore, while the derivations given in this chapter are general and can be applied to either refracting or reflecting surfaces, most of the examples discussed in this book relate to reflecting optical systems.

2.2 Useful first-order optics

Some basic ray tracing is required to produce an optical plate diagram from any given optical system comprised of spherical or aspheric lenses or mirrors. The necessary first-order tools are given in this section. We can begin with some necessary definitions.

As with Conrady [3] and Kingslake and Johnson [4], this book uses the following sign convention:

1) Right-handed Cartesian coordinates for X, Y, and Z. The Z-axis is horizontal on the page, positive in the left-to-right direction, and, unless otherwise stated, defines the optical axis of a system with rotational symmetry. The positive Y-axis is vertical in the 'up-page' direction.
2) Fans of rays in the plane of the page contain the optical axis (Z) and are therefore 'meridian plane rays'.
3) The acute angle that a meridian plane ray makes with the Z-axis is positive if the ray is rotated clockwise from the optical axis about the axis intercept point.
4) The radius of curvature of a surface is denoted by r. Its reciprocal is the curvature of the surface, $c = \frac{1}{r}$.
5) The radius is positive when its centre of curvature lies in the positive Z-direction from the surface vertex.
6) The angle between the surface normal at the ray intercept and the optical axis is φ.
7) The local origin of the Cartesian coordinate system is defined as the point where the optical axis of symmetry (Z) intersects the optical surface. This

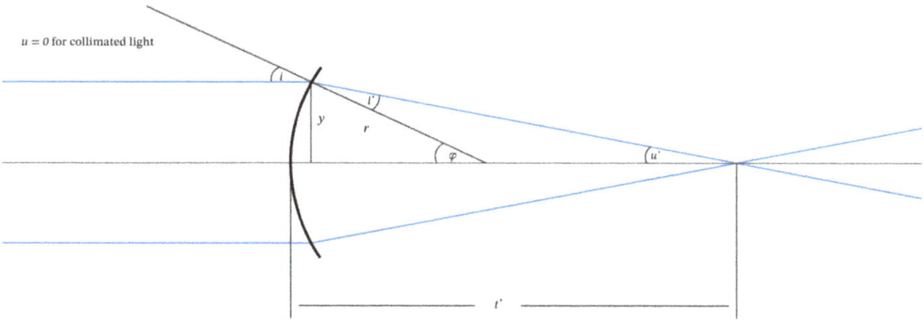

Figure 2.2. Conventions used in this book. All quantities shown in this diagram are positive.

point is the pole of the surface, and therefore the XY plane can be referred to as the polar tangent plane.

8) In sequential ray tracing, the origin is transferred surface by surface as a ray transits the system. Here, x, y and t denote object space distances (along the Z-axis) before refraction and x', y' and t' denote the distances after refraction.
9) The angle of incidence that the ray makes with respect to the surface normal is i. After refraction, the corresponding angle is i'.
10) The angle of incidence i is positive when it is produced by an anti-clockwise rotation about the surface normal.

Figure 2.2 provides a summary of this information. All quantities shown in the diagram are positive.

Recalling the paraxial version of Snell's Law (equation (2.5)), from figure 2.2 we can see that

$$ni = n'i' \rightarrow n(\varphi - u) = n'(\varphi - u'), \tag{2.6}$$

from which we obtain

$$n'u' - nu = (n' - n)\varphi = \left(\frac{n' - n}{r}\right)y = (n' - n)cy. \tag{2.7}$$

The quantity $\left(\frac{n'-n}{r}\right)$, or equivalently $(n' - n)c$, is called the optical power of the surface, denoted by K.

In ray tracing, the relationship

$$n'u' - nu = Ky \tag{2.8}$$

is used to calculate u' after refraction at a surface. This is the first of two fundamental ray-tracing relationships.

The second fundamental ray-tracing relationship takes the ray after refraction at a surface at height y, and calculates its new height, y_+, after it has travelled an axial distance of t' to the next surface.

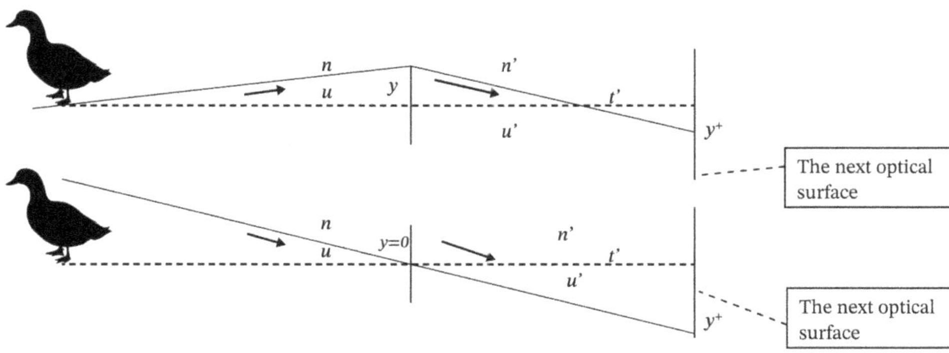

Figure 2.3. Ray diagrams showing a marginal axial ray (top) and principal ray of the most oblique pencil, or 'chief ray' (bottom), being refracted at a first surface and transferring to a second surface.

$$y_+ = y - u't'. \tag{2.9}$$

The rays traced through an optical system are usually the marginal (edge) ray of the axial pencil and the principal ray of the most oblique pencil (otherwise referred to as the chief ray). The refractive indices before and after refraction are n and n'.

Figure 2.3 shows an axial marginal ray intersecting the first surface at height y and being refracted, then travelling to the second surface, intercepting it at height y_+. The lower diagram shows a principal ray (from the most oblique pencil from the head of the duck) being refracted at the first surface at $y = 0$, then continuing on to the second surface, which it intercepts at height y_+. Knowing initial values for u and y as well as n, n' and the power of the surface at y, K, we can calculate u' and y_+ and repeat this for as many surfaces as it takes through the system.

Rearranging equation (2.8) for refraction at a surface, and considering that $y' = y$ on refraction at a surface (there is no change in y without some axial change t') we can produce the following matrix equation:

$$(nu,\ y)\begin{pmatrix} 1 & 0 \\ K & 1 \end{pmatrix} = (n'u',\ y'). \tag{2.10}$$

Taking next the transfer (equation (2.9)), we can produce:

$$(n'u',\ y')\begin{pmatrix} 1 & \dfrac{-t'}{n'} \\ 0 & 1 \end{pmatrix} = (n'u'_+,\ y_+). \tag{2.11}$$

Combining these results gives:

$$(nu,\ y)\begin{pmatrix} 1 & 0 \\ K & 1 \end{pmatrix}\begin{pmatrix} 1 & \dfrac{-t'}{n'} \\ 0 & 1 \end{pmatrix} = (n'u'_+,\ y_+). \tag{2.12}$$

This gives the ray axial angle and height from given starting values after refraction at a surface, then transfer to the next surface. Note that the angle is multiplied by the refractive index of the space it is in here.

Considering rays in the figure, note how they meet the first surface. The first surface is a pupil. A pupil is either an 'aperture stop': a physical aperture that limits the beam diameter of all pencils of rays going through the system, or it is the image of an aperture stop through one or more optical surfaces in the system. All principal rays go through the centre of the pupil, and all marginal rays go through the edge of a pupil.

We can now combine surface and transfer matrices for any number of surfaces, giving the final result:

$$(n_1u_1,\, y_1)\begin{pmatrix} 1 & 0 \\ K_1 & 1 \end{pmatrix}\begin{pmatrix} 1 & \dfrac{-t'_1}{n'_1} \\ 0 & 1 \end{pmatrix}\cdots\begin{pmatrix} 1 & 0 \\ K_j & 1 \end{pmatrix}\begin{pmatrix} 1 & \dfrac{-t'_j}{n'_j} \\ 0 & 1 \end{pmatrix} = (n_ju_j,\, y_j). \tag{2.13}$$

Equation (2.13) allows us to calculate all the necessary first-order quantities to produce a plate diagram from a system of any number of elements. Also, as we will see in the following sections, these quantities are used in the steps taken to evaluate spherical aberration at each surface in multi-element systems. Setting up matrices such as this for a multi-surface system is quite straightforward and simple to keep track of, precluding the need to write out the 'yard long equations' that Burch complains about for multi-element systems [5].

This approach does take some getting used to, and keeping track of sign is critical. It is an excellent exercise for a student of optics to become familiar with this sort of equation.

2.3 The Schmidt telescope

Before we proceed to the calculation of third-order aberration coefficients, it is useful to consider how these aberrations arise. The Schmidt telescope [6] is a very good fit for this purpose and also leads naturally to the plate diagram, as it did for Burch.

In figure 2.4(a), we see a spherical mirror with the entrance pupil plane at the centre of curvature of the mirror. The principal rays of each collimated pencil of rays intersect at the optical axis at the entrance pupil. As the entrance pupil is at the centre of a spherical mirror, each principal ray originates from the centre of the spherical surface and thus meets the mirror surface at normal incidence.

The system is completely symmetric in this sense; therefore, pencils of rays from all field points receive an equal amount of spherical aberration from the mirror; there is no 'field-varying off-axis aberration' affecting image sharpness to the third order, no coma, and no astigmatism. Spherical aberration provides the same blurring at any point in the image, which forms on a spherical surface half-way between the mirror surface and its centre of curvature.

In figure 2.4(b), a plate with a fourth-degree term is placed at the centre of curvature of the mirror, and its profile is chosen to exactly cancel the spherical aberration of the spherical mirror. As each pencil of rays only suffers from wavefront deformation due to spherical aberration and the plate is at a pupil, all

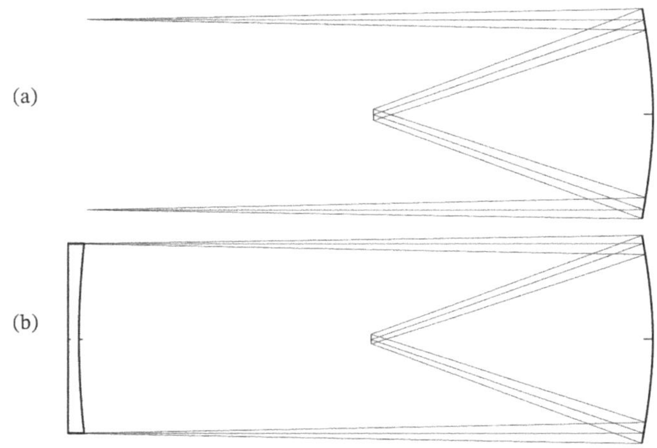

Figure 2.4. A spherical mirror with an aperture stop at the centre of curvature (a) and a Schmidt telescope (b). In the Schmidt telescope, a 4th-power wavefront retarding plate is placed at the centre of curvature of the spherical mirror, with its retardation adjusted to cancel the 4th-power advance to the wavefront that is imparted by the mirror.

field points receive the same correction from the plate, which exactly cancels the equal spherical aberration each pencil of rays receives from the mirror. The system in figure 2.4(b) is therefore simultaneously free of spherical aberration, coma, and astigmatism. Systems with such correction are called 'anastigmats'.

The third-order spherical aberration of a spherical mirror in collimated light is given by:

$$W_{\text{Mirror}} = -\frac{1}{4}c^3y^4, \tag{2.14}$$

where c is the curvature of the mirror, and y is the marginal ray height of the axial pencil of rays. This equation will be fully derived in the following section.

The Schmidt plate cancels the spherical aberration of the spherical mirror, and the spherical aberration strength of the plate to the left of figure 2.4(b) is therefore:

$$W_{\text{Plate}} = \frac{1}{4}c^3y^4. \tag{2.15}$$

Following the usual conventions, the radii as drawn in figure 2.4 are negative, so the spherical aberration of the spherical mirror is positive, meaning that marginal rays come to focus before paraxial rays. We can see in figure 2.4(b) that the plate has the opposite effect, curling the wavefront back at the edges and therefore retarding marginal rays, or 'curling back' the rim of the wavefront.

This level of description of a Schmidt telescope is sufficient for the purposes of this book. A far more thorough technical description of the Schmidt telescope is given by Linfoot [7]. In the next section, we derive the wavefront spherical aberration coefficients given in equations (2.14) and (2.15), as well as some other equivalent expressions that are used for the plate diagram.

2.4 Hopkins's *Wavefront Spherical Aberration Coefficient*

The following derivation is largely based on one given in H H Hopkins's 1950 book *Wavefront Aberration Coefficients* [8].

Spherical aberration (in wavefront terms) is usually defined as the maximum aspherical optical path difference (OPD), i.e. the differential across the wavefront introduced between the incident and transmitted wavefronts for axially symmetrical systems. This definition excludes changes in the focus of the spherical reference wavefront itself.

The spherical aberration arising at an optical surface, measured in wavefronts, is defined in terms of the maximum optical path difference introduced between the incident and transmitted wavefronts. For our purposes, the marginal ray of the axial pencil is the only ray required to quantify the third-order spherical aberration produced by a given surface.

Figure 2.5, reproduced from Hopkins, identifies quantities used in the derivation of the third-order spherical aberration coefficient. In this diagram, every defined quantity is of positive sign.

Considering figure 2.5, and assuming that the indicated ray represents a marginal ray of an axial pencil, we can begin by observing that the optical path difference introduced into this ray on refraction at the surface can be expressed as:

$$W = n'(L' - pO') - n(L - pO). \tag{2.16}$$

We can see that

$$\begin{aligned} pO^2 &= y^2 + (L - z)^2, \\ &= y^2 + L^2 - 2Lz + z^2. \end{aligned} \tag{2.17}$$

Also,

$$\chi^2 = y^2 + z^2. \tag{2.18}$$

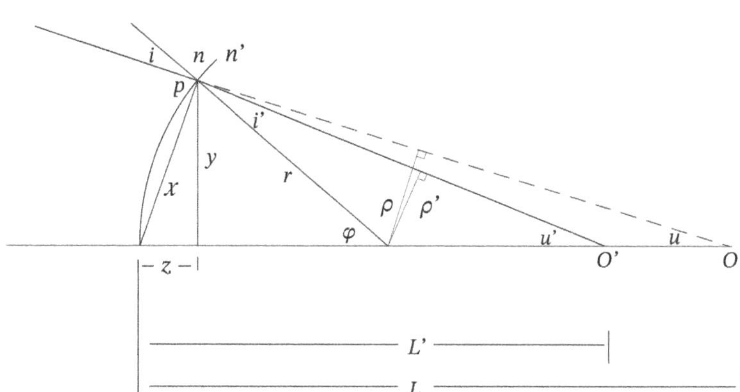

Figure 2.5. Quantities used in deriving the third-order wavefront spherical aberration coefficient. All quantities shown in this diagram are positive.

For a spherical surface, the surface profile equation is

$$y^2 = 2rz - z^2,$$

from which, upon substituting equation (2.18), we obtain

$$2rz = \chi^2 \rightarrow 2z = \frac{\chi^2}{r}. \tag{2.19}$$

Combining equations (2.17) and (2.19) gives

$$\begin{aligned} pO^2 &= L^2 - \frac{L\chi^2}{r} + \chi^2 \\ &= L^2\left(1 - \frac{\chi^2}{Lr} + \frac{\chi^2}{L^2}\right) \\ &= L^2\left(1 - \frac{\chi^2}{L}\left(\frac{1}{r} - \frac{1}{L}\right)\right). \end{aligned} \tag{2.20}$$

Using the binomial theorem,

$$pO = L(1 - \frac{\chi^2}{2L}\left(\frac{1}{r} - \frac{1}{L}\right) - \frac{\chi^4}{8L^2}\left(\frac{1}{r} - \frac{1}{L}\right)^2 + O^6. \tag{2.21}$$

Excluding high-order terms and substituting equation (2.21) into the terms on the rhs of equation (2.16) gives

$$\begin{aligned} n(L - pO) &= \frac{1}{2}\chi^2 n\left(\frac{1}{r} - \frac{1}{L}\right) + \frac{1}{8}\chi^4\left(n\left(\frac{1}{r} - \frac{1}{L}\right)\right)^2 \frac{1}{nL} \\ n'(L' - pO') &= \frac{1}{2}\chi^2 n'\left(\frac{1}{r} - \frac{1}{L'}\right) + \frac{1}{8}\chi^4\left(n'\left(\frac{1}{r} - \frac{1}{L'}\right)\right)^2 \frac{1}{n'L'}. \end{aligned} \tag{2.22}$$

Observing that

$$\varphi = \frac{y}{r} \text{ and } u = \frac{y}{L}$$

leads to

$$\begin{aligned} i &= \varphi - u = \left(\frac{y}{r} - \frac{y}{L}\right), \\ i' &= \varphi - u' = \left(\frac{y}{r} - \frac{y}{L'}\right). \end{aligned} \tag{2.23}$$

We see that in the paraxial limit, Snell's law can be written as

$$n\left(\frac{y}{r} - \frac{y}{L}\right) = n'\left(\frac{y}{r} - \frac{y}{L'}\right). \tag{2.24}$$

Also, in the paraxial limit,

$$\chi \rightarrow y.$$

Finally, applying these paraxial approximations to equation (2.22), substituting into equation (2.16), and disregarding high-order terms, we can write the equation for the third-order change in the OPD of the marginal paraxial ray as

$$\begin{aligned} W &= \frac{1}{8}y^4\left(n\left(\frac{1}{r} - \frac{1}{L}\right)\right)^2\left(\frac{1}{n'L'} - \frac{1}{nL}\right) \\ &= \frac{1}{8}y\left(n\left(\frac{y}{r} - \frac{y}{L}\right)\right)^2\left(\frac{y}{n'L'} - \frac{y}{nL}\right) \\ &= \frac{1}{8}y(ni)^2\left(\frac{u'}{n'} - \frac{u}{n}\right). \end{aligned} \tag{2.25}$$

Equation (2.25) gives the third-order spherical aberration for refraction at a surface, measured in wavefronts. Two other variants of this expression are useful in the construction of plate diagrams for reflecting systems in air.

In this case, where $n = \pm 1$ and $n' = \mp 1$, then $n^2 = n'^2 = 1$, $n^3 = \pm 1$, and $n'^3 = \mp 1$.

First, using quantities defined earlier, we have

$$u + i = u' + i' = \varphi = \frac{y}{r} = cy. \tag{2.26}$$

Also

$$\begin{aligned} \left(\frac{u'}{n'} - \frac{u}{n}\right) &= n'^2\frac{u'}{n'} - n^2\frac{u}{n} \\ &= n'u' - nu \\ &= n'(u' + i') - n(u + i) \\ &= y\left(\frac{n' - n}{r}\right) \text{ or } y(n' - n)c \\ &= -2nyc. \end{aligned} \tag{2.27}$$

Substituting this into equation (2.25) gives

$$\begin{aligned} W &= \frac{1}{8}y(ni)^2\left(\frac{u'}{n'} - \frac{u}{n}\right) \\ &= -\frac{1}{4}y(ni)^2(-2nyc) \\ &= -\frac{1}{4}n^3i^2y^2c \\ &= -\frac{1}{4}nci^2y^2. \end{aligned} \tag{2.28}$$

Surface#	Surface Name	Surface Type	Y Radius	Thickness	Glass	Refract Mode	Y Semi-Aperture
Object		Sphere	Infinity	Infinity		Refract	
Stop		Sphere	-500	-250		Reflect	50
Image		Sphere	Infinity	0.0000		Refract	4.7708
End Of Data							

THIRD ORDER IMAGE ABERRATIONS (Wavelengths at 1000 nm)					
Surface	W040	W131	W220	W222	W311
SUM	12.5000	-8.7275	-0.0000	1.5234	-0.0000

Figure 2.6. Confirmation of equation (2.29) in Code V software. The W040 column gives the third-order wavefront spherical aberration coefficient.

Second, we note that in collimated light:

$$u \to 0 \text{ and } i = \frac{y}{r} = cy.$$

Substituting this into equation (2.28) gives:

$$W = -\frac{1}{4}nc^3y^4. \tag{2.29}$$

It is considered useful to do a quick check of the math against a software tool at this point. For a worked example, consider a ray at a spherical mirror in air with the following parameter values:

$$r = -500 \text{ mm}; \; c = -\frac{1}{500}\text{mm}^{-1},$$

$$y = 50.0 \text{ mm},$$

$$W = -\frac{1}{4}\left(-\frac{1}{500}\right)^3 50^4 \text{ mm} = 0.0125 \text{ mm}.$$

Inserting these values into the Synopsys Code V software, we get the result in figure 2.6.

We next use equation (2.29) and a conceptual approach that is useful for the plate diagram to derive the other four third-order aberration coefficients.

2.5 Off-axis third-order aberration

Recalling section 2.3 of this chapter, we saw how a Schmidt plate exactly cancelled the spherical aberration of a spherical mirror in collimated incident light. In the case of a Schmidt telescope, the plate is placed at the aperture stop, which is located at the centre of curvature of the spherical mirror. This ensures that the principal ray of any pencil of collimated rays, passing through the centre of the aperture stop, which is also the centre of curvature of the mirror, strikes the mirror at normal incidence.

Therefore, all pencils of rays encounter exactly the same spherical reflecting surface, and the mirror gives exactly the same amount of spherical aberration to each pencil of rays. As all pencils of rays exactly overlap at the entrance pupil, the Schmidt plate placed here gives the exact same correction to every pencil, to the third order.

Burch realised that this implied that the third-order aberrations generated in a wavefront on reflection at the spherical mirror could be replicated entirely by passing the wavefront through an 'anti-Schmidt' plate. For flat wavefronts, this plate imparts exactly the same magnitude of optical path difference (but with the opposite sign) as the Schmidt plate, with a 4th-degree radial dependence. Figure 2.7 extends figure 2.4 above to illustrate this concept.

The anti-Schmidt plate that remains in figure 2.7(d) gives rise to exactly the same third-order aberrations as the original spherical mirror did. This remains true as the aperture stop is moved away from the centre of curvature of the mirror and asymmetrical off-axis aberrations evolve. However, the plate performs no focusing (first-order change); light remains collimated before and after the plate.

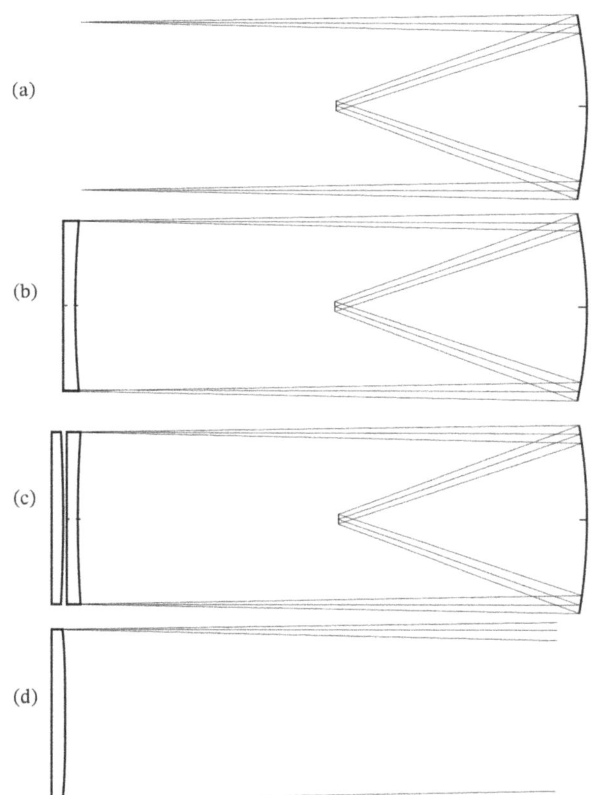

Figure 2.7. (a) A spherical mirror with a stop at the centre of curvature, (b) a Schmidt camera with an aspheric plate at the stop to correct spherical aberration, (c) a second plate is added to the Schmidt plate, which has the same spherical aberration strength as the spherical mirror, and (d) removing the anastigmatic Schmidt camera from (c), we are left with a zero-power optical system that reproduces all the third-order behaviour of the original spherical mirror. This is the plate diagram for a spherical mirror.

This is the key realisation behind the optical plate diagram. Considering now the change in optical path distance in a collimated pencil of off-axis rays that pass through the anti-Schmidt plate when it is not at the aperture stop, we have a means by which we can calculate the off-axis aberrations of the spherical mirror.

Figure 2.8 illustrates such a situation.

The retardance of the plate in figure 2.8 decreases as the 4th power of the zonal radius. A collimated pencil of non-axial rays, making an angle u with respect to the optical axis, is incident on the plate at an axial distance x from the stop. As the light is collimated through the stop, the height of the marginal ray of the axial pencil at the plate is the semi-diameter of the aperture stop, y_c. The principal ray of the meridional fan of rays in the plane of the page strikes the plate at height y_p.

The optical axis is normal to the plate through its centre of symmetry. The oblique pencil crossing the plate has a circular cross section with a radius y_c. The principal ray intersects the plate at the centre of this cross section at height y_p. In the cross section of the oblique pencil, we set up polar coordinates $(\rho, \ \theta)$, with the origin on the principal ray, $\theta = 0$ for the vertical upwards direction, and with ρ ranging from zero to y_c. X and Y give Cartesian coordinates in the plane of $(\rho, \ \theta)$ but with the origin on the optical axis, as illustrated in figure 2.9.

This plate will advance or retard a wavefront in proportion to the 4th power of the distance from the optical axis. A constant of proportionality, κ, can be introduced, and its value is set such that $W = \kappa y_c^4$ or $\kappa = \frac{W}{y_c^4}$.

From the diagram, we have

$$X = \rho \sin \theta$$

$$Y = y_p + \rho \cos \theta.$$

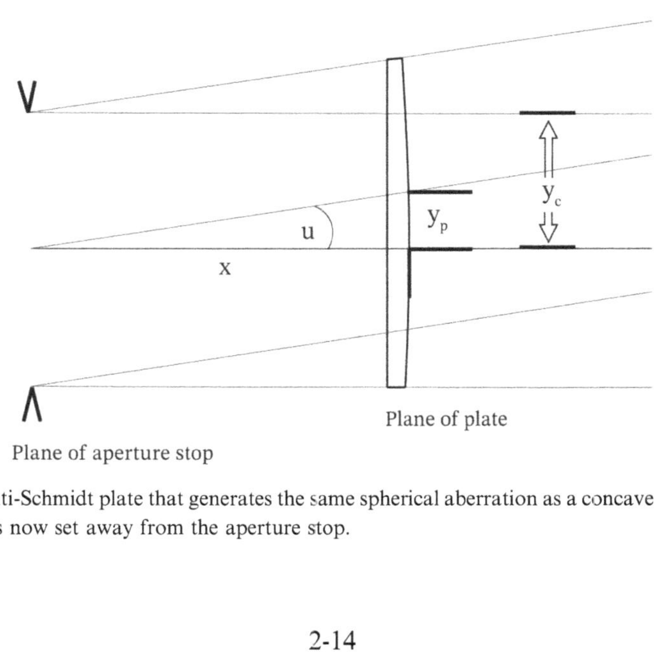

Figure 2.8. An anti-Schmidt plate that generates the same spherical aberration as a concave spherical mirror in collimated light is now set away from the aperture stop.

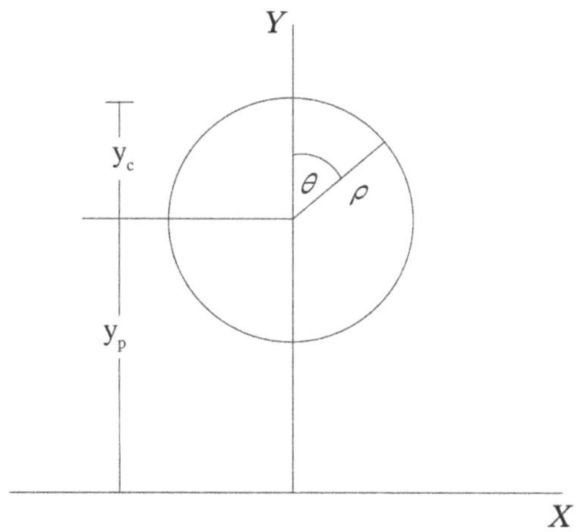

Figure 2.9. Quantities used in the derivation of off-axis aberrations. The XY plane is in the plane of the plate, and the origin is on the optical axis. Polar coordinates are defined on the cross section of the oblique pencil of rays in the plane.

The advance or retardation of the wavefront, δW, for any point on the plate is then

$$\begin{aligned}\kappa(X^2 + Y^2)^2 &= \kappa\left((\rho \sin \theta)^2 + \left(y_p + \rho \cos \theta\right)^2\right)^2 \\ &= \kappa\left(\rho^2\sin^2\theta + y_p^2 + 2y_p\rho \cos \theta + \rho^2\cos^2\theta\right)^2 \\ &= \kappa\left(y_p^2 + 2y_p\rho \cos \theta + \rho^2\right)^2 \\ &= \kappa\left(y_p^4 + 4y_p^3\rho \cos \theta + 2y_p^2\rho^2 + 4y_p^2\rho^2\cos^2\theta + 4y_p\rho^3\cos^3\theta + \rho^4\right).\end{aligned} \tag{2.30}$$

These terms represent, from left to right, a constant or 'piston' term, distortion, curvature, astigmatism, coma, and spherical aberration.

Equation (2.30) shows the dependence of third-order aberrations on y_p, ρ and θ. Commonly, we wish to have a single number for each aberration. Typically, maximal values are used for y_p and ρ, and θ is set to zero.

Equation (2.30) contains four terms that have field dependence; that is, they are dependent on the height of the principal ray of an off-axis pencil of rays at the plate, y_p. As we see in figure 2.10, the field angle remains constant through a system of any number of plates, so if we take the distance from the entrance pupil to the ith plate to be x_i, then

$$y_{pi} = -ux_i. \tag{2.31}$$

Using this result, substituting $\frac{W}{y_c^4}$ for κ, y_c for ρ (as it is the maximum value of ρ), and with θ set to zero, we can obtain the following five expressions for third-order aberrations from equation (2.30):

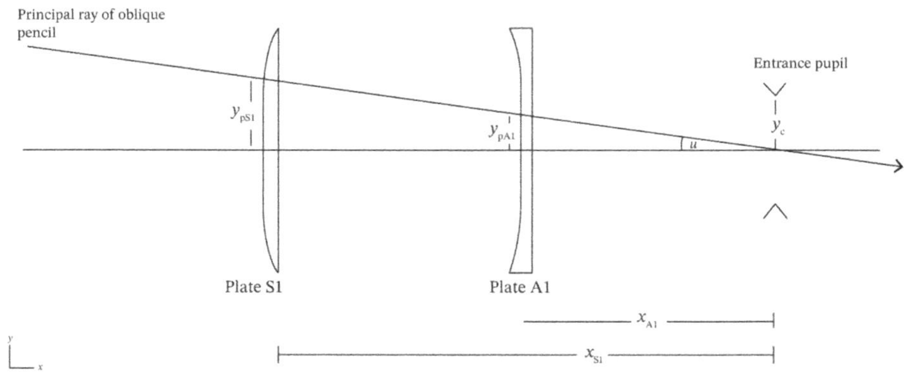

Figure 2.10. Two plates are shown displaced at distances x_{A1} and x_{S1} from the system entrance pupil. The field angle u remains invariant throughout the system of plates, so the height of the ray at the plate is the product of a system constant u and x_i, the distance of the ith plate from the entrance pupil.

$$
\begin{aligned}
&\frac{W}{y_c^4}(\rho^4) &&\rightarrow \frac{W}{y_c^4}\left(y_c^4\right) &&= W && \text{spherical aberration}\\
&\frac{W}{y_c^4}(4y_p\rho^3\cos^3\theta) &&\rightarrow -4\frac{W}{y_c^4}ux\left(y_c^3\right) &&= -4\frac{u}{y_c}xW && \text{coma}\\
&\frac{W}{y_c^4}\left(4y_p^2\rho^2\cos^2\theta\right) &&\rightarrow 4\frac{W}{y_c^4}(ux)^2\left(y_c^2\right) &&= 4\left(\frac{u}{y_c}\right)^2 x^2W && \text{astigmatism}\\
&\frac{W}{y_c^4}\left(2y_p^2\rho^2\right) &&\rightarrow 2\frac{W}{y_c^4}(ux)^2\left(y_c^2\right) &&= 2\left(\frac{u}{y_c}\right)^2 x^2W && \text{field curvature}\\
&\frac{W}{y_c^4}(4y_p\rho^3\cos\theta) &&\rightarrow -4\frac{W}{y_c^4}(ux)^3\left(y_c^3\right) &&= -4\left(\frac{u}{y_c}\right)^3 x^3W && \text{distortion}
\end{aligned}
\tag{2.32}
$$

Figure 2.10 introduces the idea that in 'plate space', angles are system constants to the first order. The only terms in equation (2.32) that are specific to each plate are X and W. Figure 2.11 shows the wavefront shapes resulting from the aberrations in equation (2.32).

- **Distortion** manifests as a radial wavefront tilt increasing cubically with field angle. This causes a displacement of image points radially outward (or inward), proportional to the cube of the field angle, without affecting focus or blur size.
- **Field curvature** generates a wavefront with uniform curvature over the pupil, increasing quadratically with field angle. The result is that all image points nominally lie on a curved surface (to the third order, a sphere), rather than a flat image plane.
- **Astigmatism** produces a wavefront where one half of the pupil is advanced and the other retarded, causing fans of rays in the X and Y directions to focus on axially separated surfaces and produce two lines of best focus at right angles to each other, rather than a focal point. In other words, the image

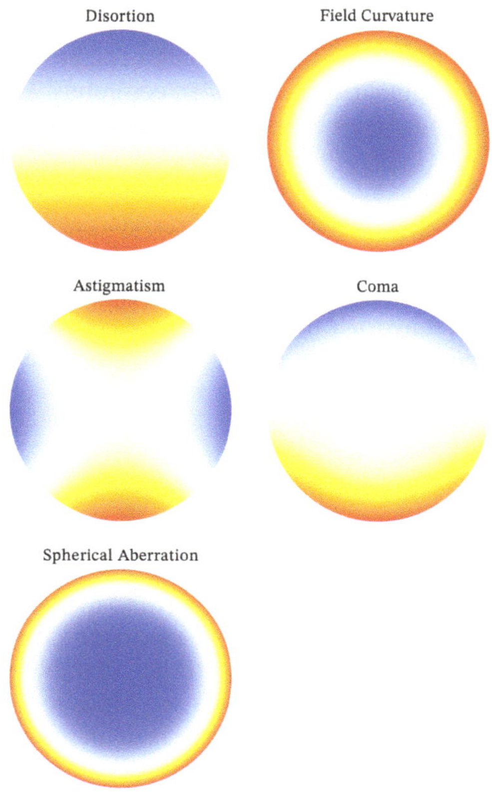

Figure 2.11. Graphical representation of the wavefront shapes arising from the five aberration terms in equation (2.30).

of a point object forms a line focus in one orientation, and then another line at right angles to the first and further along the optical axis. The curved focal surface arising from field curvature lies midway between these two focal lines. In this surface, images are circular, but blurred to a diameter of half the length of the astigmatic focal lines.

- **Coma** leads to an asymmetric wavefront skewed along the field angle. Considering wavefront rays from semicircular regions above and below the pupil centre, each forms overlapping circular blur patches at different lateral positions in the image. The combined result is a triangular comet-like blur, with an apex angle of 60 degrees.
- **Spherical aberration** is independent of field angle and introduces a symmetric wavefront deviation growing as the fourth power of the pupil coordinate, leading to peripheral rays focusing closer or further than paraxial rays, depending on the sign.

Most commonly, the distortion term has been neglected in plate diagram analysis. Gascoigne first derived it, and Linfoot presented this in *Recent Advances in Optics* (see reference [7]). Distortion does not affect the sharpness of point images, only the mapping. Aplanatic systems (simultaneously corrected for spherical aberration and coma) or anastigmatic optical systems (simultaneously corrected for spherical aberration, coma, and astigmatism) are most commonly sought after, as point blurring limits system sensitivity, whereas mapping errors can be dealt with in post-processing.

The 'curvature' term, with no θ dependence, is tied to astigmatism through its shared y_p and ρ quadratic dependence. Together, these terms define a surface of best focus in the presence of astigmatism. There is one final term, related to these, that does not appear in the wavefront expansion: Petzval curvature. If an optical system of real lens elements and mirrors forms an image free of astigmatism (and therefore also field curvature), the images form on the Petzval surface.

The Petzval surface is spherical and has a curvature given by:

$$c_p = \sum_{i=1}^{m} \frac{n_i' - n_i}{n_i' n_i r_i}, \text{ which for mirrors in air} \rightarrow \sum_{i=1}^{m} \frac{2}{r_i}. \tag{2.33}$$

Petzval curvature stands apart from the wavefront aberration coefficients in that it has no dependence on y_p, ρ or θ. Only the curvatures of the real optical elements contribute to it. If the sum in equation (2.33) is equal to zero, the Petzval surface is flat. But for the image to be flat in this case, the field curvature term (and therefore the astigmatism) must also be zero; otherwise, circular blurred images will form exactly between the two astigmatic focal surfaces. An easy way to see how Petzval curvature arises, attributable to Dr John Rogers, is to imagine principal rays fanning out from the centre of curvature of the spherical mirror in a Schmidt telescope [9]. Each ray hits the mirror at normal incidence and reflects straight back on itself, while the mirror brings the parallel rays from the pencil surrounding each principal ray to a sharp focus at half the radius of the mirror. The locus of focus points thus formed defines a spherical surface concentric with the spherical mirror and having half its radius. This is the Petzval surface.

Figure 2.12 summarises the relationship between tangential, mean, sagittal, and Petzval focal surfaces.

Sometimes, astigmatism is deliberately used to 'correct' Petzval curvature. The Petzval lens referred to in chapter 1 did exactly that. A fixed amount of astigmatism is deliberately introduced so that, say, the mean focal surface, with its circular but blurred images, is flat. Or one of the astigmatic focal surfaces can be driven flat, typically the tangential surface, as that minimises the total amount of astigmatism used to correct the curvature. In the later part of the 19th century, Charles Piazzi Smyth (the first person to prove Newton's theory that astronomical images would be more sharply formed when taken from the peaks of high mountains) was interested in taking wide-angle images of hieroglyphs in the Great Pyramid at Giza. He realised that he could improve image sharpness if he completely corrected astigmatism and introduced a strongly curved lens close to the focal surface to correct the Petzval sum to zero. This lens has little impact on aberration; because it is close to the focal surface, the marginal ray height at its curved surface is almost zero.

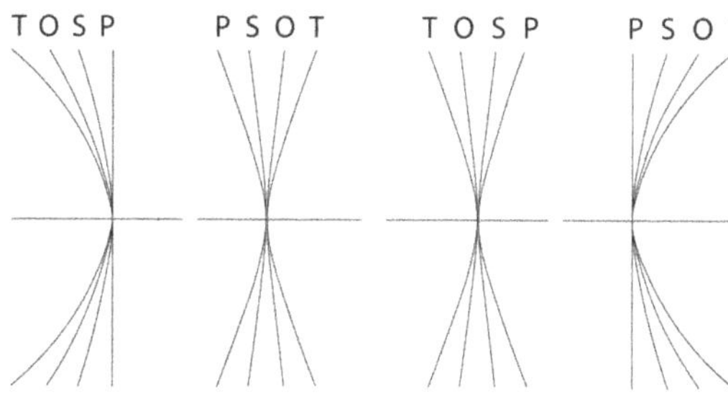

P = Petzval Surface
O = Mean Image Surface
S = Sagittal Focal Surface
T = Tangential Focal Surface

Figure 2.12. Drawing showing the various focal surfaces arising from Petzval curvature and astigmatism. In the far left and far right plots, the Petzval surface is flat and astigmatism has opposite signs. The inner two plots show how nonzero Petzval curvature can make the mean image surface flatter. The horizontal (axis-parallel) distances from the Petzval surface to the sagittal, mean, and tangential focal surfaces, respectively, are always in the ratio 1:2:3.

According to Kingslake, Piazzi Smyth remarked in 1874, they 'relieved us of a blunder by substituting a sin' [10].

Typically, a flat image surface is desirable, as most image-forming formats, such as photographic film, or electronic detectors, are flat. While the plate diagram does not include Petzval curvature, the functional form of Petzval curvature in terms of lens constructional parameters is so simple that including 'flat-field' constraints in a system under plate diagram analysis is trivial.

2.6 Single-element plate diagram

Using the tools developed above, we shall begin our plate diagram investigation with a simple system, namely a paraboloid mirror in collimated light. Following Burch, we can think of the mirror as having two parts contributing to aberration, each of which can be represented by two Schmidt-like plates [11].

The spherical aberration of the paraboloid mirror in collimated light is exactly zero. It has been known since ancient times that the conic section cut with an axis parallel to the side of the cone has two perfect foci: one at half the value of the vertex radius of curvature and the other at infinity. This was pointed out by Diocles in ~200 B.C.E. [12].

This means that the departure from sphericity of the mirror, which makes the mirror a paraboloid, must be represented by a plate that exactly cancels the spherical aberration of the spherical mirror and therefore must be identical to the Schmidt plate that would correct the base spherical mirror.

In the Schmidt telescope, the plate is colocated with the astigmatising plate that represents the spherical mirror, at the mirror's centre of curvature. The plates exactly cancel each other, creating an essentially flat window and yielding an aberration-free system.

In the case of the paraboloid mirror, the plate is located at the mirror surface. This is a fundamental difference between a Schmidt telescope and a paraboloid mirror. In both cases, the plate strength is identical, to cancel the spherical aberration of the base sphere.

With the paraboloid, the two plates are axially separated by the radius of curvature of the mirror, and while spherical aberration cancels, coma cancellation is impossible, and astigmatism cancellation requires one extra step, as will be shown.

The spherical aberration of the parabolising plate exactly cancels the spherical aberration of the base sphere. If we define W_S as the plate giving the spherical aberration of the base sphere and W_A as the plate giving the spherical aberration associated with the asphericity of the mirror, in this case, parabolisation, we can write

$$W_A = -W_S. \tag{2.34}$$

A three-dimensional conicoid shape is generated by rotating a two-dimensional conic section curve about its axis. If we consider the eccentricity of the generating conic, in this case a parabola, to be e, then the conic constant,[1] k, is $-e^2$. In the case of a parabola, $k = -1$. The spherical aberration for a conicoid asphericity, apart from the special case of a paraboloid, can be written as

$$W_A = -k\frac{1}{4}nc^3y^4. \tag{2.35}$$

Equation (2.35) remains true for any conicoid shape in any part of a multi-surface optical system, whether the light is diverging, converging, or collimated. But W_S must be calculated using equation (2.28) in converging or diverging light, and equation (2.29) is an alternative form that is correct only in collimated light.

In the case we are considering, k has been assigned a value of -1, giving a paraboloid; however, more generally, k can represent any of the possible conicoids, depending on the following ranges:

$k < -1$	hyperboloid, generated by rotating a hyperbola about its axis;
$k = -1$	paraboloid, generated by rotating a parabola about its axis;
$-1 < k < 0$	ellipsoid, generated by rotating an ellipse about its major axis;
$k = 0$	sphere, generated by rotating a circle about its diameter;
$k > 0$	oblate spheroid, generated by rotating an ellipse about its minor axis.

With this result, we can now look at equation (2.32) and list the terms for spherical aberration, coma, and astigmatism that apply to the paraboloid mirror:

$$W_S + W_A \rightarrow W_S - W_S = \text{spherical aberration},$$

[1] The conic constant was first used by Karl Schwarzschild in his masterly derivation of aberrations to the 5th order from the Eikonal and his application of these to the design of two-mirror telescopes, which will be discussed in the next section [13, 14].

$$-4\frac{u}{y_c}W_S(x_S - x_A) = \text{coma},$$

$$4\left(\frac{u}{y_c}\right)^2 W_S\left(x_S^2 - x_A^2\right) = \text{astigmatism}. \tag{2.36}$$

We can see here that, as Aldis pointed out 125 years ago [15], the coma and astigmatism terms can be seen as equivalent to a system of parallel forces acting on a bar at some distance from a pivot point. In this analogy, the spherical aberration sum represents the zeroth moment of the system, the coma sum represents the first moment, and the astigmatism sum represents the second moment. To make coma or astigmatism zero, the terms $(x_S - x_A)$ and $(x_S^2 - x_A^2)$ must be driven to zero, respectively. It is immediately obvious that whatever constant value Δ we choose to add to the x values in parentheses, representing a stop shift, we cannot change the value for coma, because

$$((x_S + \Delta) - (x_A + \Delta)) = (x_S - x_A).$$

Conversely, with astigmatism, we have

$$((x_S + \Delta)^2 - (x_A + \Delta)^2) = \left(\left(x_S^2 - x_A^2\right) + 2\Delta(x_S - x_A)\right).$$

Equating the rhs to zero and solving for Δ gives

$$\Delta = -\frac{(x_S + x_A)}{2}. \tag{2.37}$$

The simple interpretation of this for any starting position of the stop is that given initial values of x_S and x_A, the change in stop position Δ that cancels the astigmatism of a paraboloid mirror locates the stop exactly midway between the mirror and its centre of curvature, i.e. at its axial focus position.

To conclude this chapter, we present a worked example that proves the equations derived so far. Table 2.1 below gives the basic parameters for a paraboloid mirror.

We choose an f-ratio of f/2.5 and a field angle of 0.1 degrees so that the third-order aberrations dominate, with only small high-order residuals. We can, for convenience, choose the wavelength to be 1 micron, so one wave is 1/1000 of a

Table 2.1. Parameters used in the calculation of a plate diagram for a paraboloid mirror.

Quantity	Value
Radius of curvature	−1000 mm
Conic constant	−1
Entrance pupil diameter	200 mm
F-number	2.5
Field of view radius	−0.1 degrees, ($\frac{-\pi}{1800}$ radians)
Initial position of aperture stop	Mirror centre of curvature

millimetre, which is useful when comparing calculated values to the output from a ray-tracing program. Alternatively, we can choose a wavelength of 1 mm, again for convenience, which requires no scaling. For a mirror calculation, we can choose any wavelength that is convenient.

Figure 2.13 shows the ray paths of the resultant system. Figure 2.14 shows the spot diagram. Clearly, on axis, the image is 'perfect' and degrades due to coma as we go off axis. Figure 2.15 shows the calculated values for third-order spherical aberration (W_{040}), coma (W_{131}), astigmatism (W_{222}).

We begin by calculating the spherical aberration of the base sphere, using equation (2.29), for a spherical mirror in collimated incident light:

$$W_{\text{sphere}} = -\frac{1}{4}nc^3y^4 = -\frac{1}{4}\left(\frac{1}{-1000\text{ mm}}\right)^3 100\text{ mm}^4 = 0.025\text{ mm}.$$

With this result, we can now evaluate coma and astigmatism for the two-plate system. With the chosen stop position at the centre of curvature of the mirror,

$$x_S = 0;\ x_A = +1000\text{ mm}.$$

Substituting these values into equation (2.36), along with other defined parameters, and using the radian value for u gives us expressions that we can evaluate for coma and astigmatism. First, for coma, we get

$$-4\frac{u}{y_c}W_S(x_S - x_A) \rightarrow -4\frac{0.0017453}{100\text{ mm}}0.025\text{ mm}(0 - 1000\text{ mm}) = -1.7453 \times 10^{-3}\text{ mm},$$

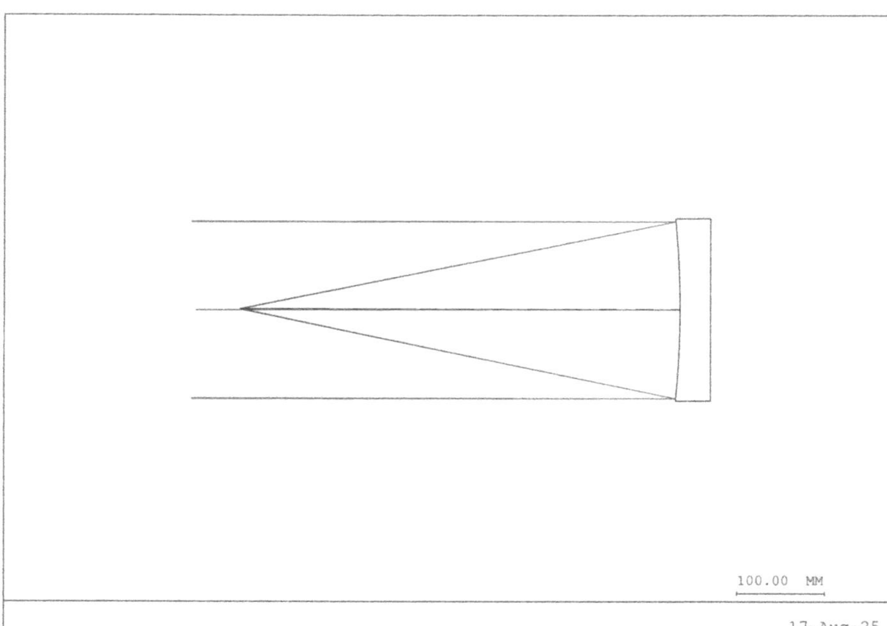

Figure 2.13. Layout diagram for the paraboloid mirror described in table 2.1. Created with Code V.

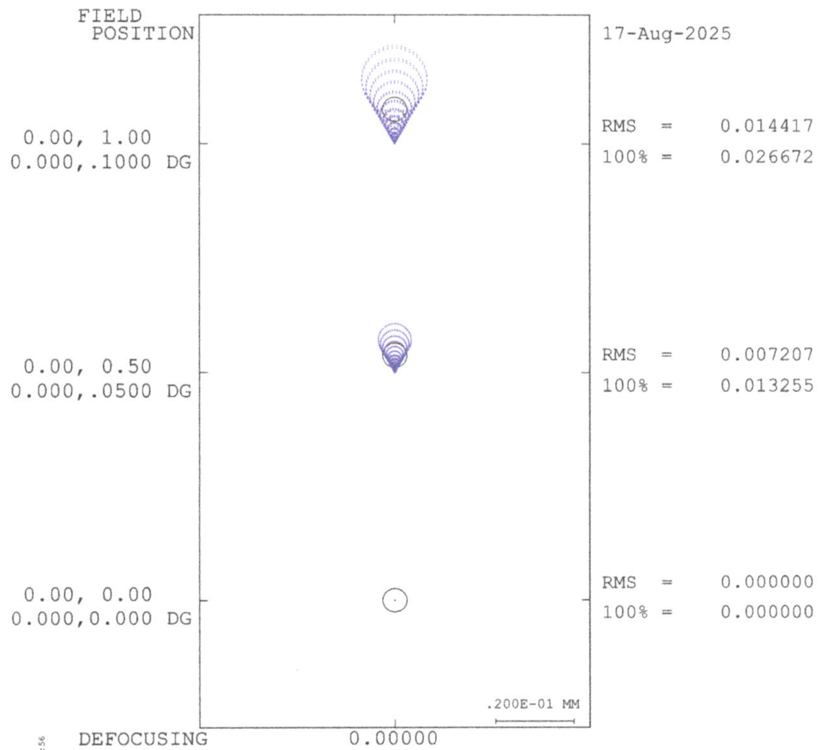

Figure 2.14. Spot diagram for the paraboloid mirror. Note that when the field doubles, the diameter of the coma patch doubles, which is consistent with the linear dependence on y_p shown in equation (2.30). Created with Code V.

and for astigmatism, we get

$$4\left(\frac{u}{y_c}\right)^2 W_S\left(x_S^2 - x_A^2\right) \rightarrow 4\left(\frac{0.0017453}{100}\right)^2 0.025\text{ mm} \times -10^6 = -3.0462 \times 10^{-5}\text{ mm}.$$

These values are checked using ray-tracing software, as shown in figure 2.16. Here, the mirror is set as a sphere.

We see that the calculated values are verified.

Next, we move the stop to the position calculated to make astigmatism zero (at the axial position of focus of the paraboloid in object space). In this case, our values for x_S and x_A become -500 mm and $+500$ mm, respectively, and

$$4\left(\frac{u}{y_c}\right)^2 W_S\left(x_S^2 - x_A^2\right) \rightarrow 4\left(\frac{u}{y_c}\right)^2 W_S \times 0 = 0,$$

as expected.

Setting the stop in the ray-tracing software gives the expected result (figure 2.17).

INFINITE CONJUGATES	
EFL	-500
BFL	-500
FFL	-500
FNO	2.5
IMG DIS	-500
OAL	0
PARAXIAL IMAGE	
HT	0.8727
ANG	0.1
ENTRANCE PUPIL	
DIA	200
THI	0
EXIT PUPIL	
DIA	200
THI	0

THIRD ORDER IMAGE ABERRATIONS (Waves at 1000 nm.)					
Surface	W040	W131	W220	W222	W311
SUM	0	-1.7453	0	0.0305	0

Figure 2.15. Calculations for focal length (EFFL), spherical aberration (W_{040}), coma, (W_{131}), astigmatism (W_{222}). The units are waves and the wavelength is 1 micron. Created with Code V.

THIRD ORDER IMAGE ABERRATIONS (Waves at 1000 nm.)					
Surface	W040	W131	W220	W222	W311
SUM	25.0000	1.7453	0	0.0305	0.0000

Figure 2.16. Wavefront third-order aberrations for the parabola with the stop at the centre of curvature: spherical aberration, coma, and astigmatism. The units of this plot are waves with a wavelength of 1 micron. Created with Code V.

THIRD ORDER IMAGE ABERRATIONS (Waves at 1000 nm.)					
Surface	W040	W131	W220	W222	W311
SUM	-0.0000	-1.7453	-0.0152	0.0000	0.0001

Figure 2.17. With the entrance pupil set exactly at the axial position of the focus of the paraboloid, coma and spherical aberration remain unchanged, while astigmatism is now zeroed. Created with Code V.

As a final step, we present a diagrammatic representation of these calculations. We can take the simplifying step of normalising the spherical aberration values to one particular value without affecting the overall balance of the equations. This is useful for simplifying the diagram without losing accuracy in terms of the overall balance between the aberrations. For exact values, the unnormalised W_i must be used (figure 2.18).

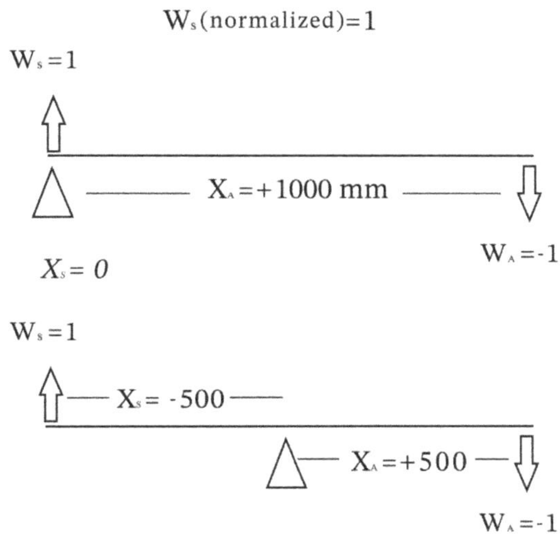

Figure 2.18. Plate diagram for a single paraboloid mirror, with the aperture stop set at the centre of curvature of the mirror (top) and midway between the centre of curvature and the surface vertex (bottom).

This is a natural point at which to introduce a theorem that will be used extensively throughout this text, the so-called 'stop-shift theorem'. Any student of optics would probably have already heard of this. In essence, using the stop-shift theorem, to the third order at least, we can rank the aberrations in terms of their field dependence. This ranking gives us the ordering of aberrations we see in equation (2.32), with spherical aberration having no dependence on field angle, coma having a linear dependence, astigmatism and field curvature having a quadratic dependence, and distortion having a cubic dependence. The stop-shift theorem states that any axial shift of the aperture stop gives rise to changes in the aberration of the field-dependent terms that are proportional to the magnitude of the stop shift and to the amount of 'lower-ranked' aberration in the system.

Equation (2.32) shows that, for a fixed value of field angle u, the field dependence of the different aberrations depends on the term y_p, the height above the axis at which the chief ray intersects the plate. Also, y_p can be expressed as $-ux$. Changing the position of an aperture stop anywhere in a multi-element system changes the position of the entrance pupil, from which all x_i are measured, so it is clear that the magnitude of the field-dependent aberrations depends on the position of the stop. Our two-plate example with the paraboloid also confirms that third-order contributions simply sum to give the system aberration. Taking constant terms outside the sum, we can express the system sum as

$$\begin{aligned}
\sum W_i &= \text{spherical aberration} \\
4\frac{u}{y_c}\sum x_i W_i &= \text{coma} \\
-4\left(\frac{u}{y_c}\right)^2\sum \frac{u}{y_c} x_i^2 W_i &= \text{astigmatism} \\
2\left(\frac{u}{y_c}\right)^2\sum x_i^2 W_i &= \text{field curvature} \\
-4\left(\frac{u}{y_c}\right)^3\sum x_i^3 W_i &= \text{distortion.}
\end{aligned} \tag{2.38}$$

Moving the aperture stop axially anywhere inside the system leads to a change in the axial location of the entrance pupil, the image of the aperture stop in object space. If this axial pupil shift is denoted by Δ_{EP}, we can reformulate equation (2.38) to include this term and evaluate the resultant change in aberrations:

$$\begin{aligned}
\sum W_i &= \text{spherical aberration} \\
4\frac{u}{y_c}\sum (x_i - \Delta_{EP}) W_i &= \text{coma} \\
-4\left(\frac{u}{y_c}\right)^2\sum (x_i - \Delta_{EP})^2 W_i &= \text{astigmatism} \\
2\left(\frac{u}{y_c}\right)^2\sum (x_i - \Delta_{EP})^2 W_i W_i &= \text{field curvature} \\
-4\left(\frac{u}{y_c}\right)^3\sum (x_i - \Delta_{EP})^3 W_i &= \text{distortion.}
\end{aligned} \tag{2.39}$$

Starting with the system sum for coma (and disregarding constant terms for clarity), we get

$$\sum (x_i + \Delta_{EP}) W_i = \sum x_i W_i - \Delta_{EP} \sum W_i. \tag{2.40}$$

The rhs shows that the system sum for coma with an axially displaced entrance pupil is equal to the original system sum for coma plus a new amount that is linearly dependent on the magnitude of the entrance pupil linear shift and on the system sum for spherical aberration. Therefore, we see that the aberration sum for the aberration that has a linear field dependence (coma) will have a new term that is dependent on the aberration sum with a zeroth-power field dependence (spherical aberration).

Repeating for astigmatism, we obtain

$$\sum (x_i - \Delta_{EP})^2 W_i = \sum x_i^2 W_i - 2\Delta_{EP}\sum x_i W_i + \Delta_{EP}^2 \sum W_i. \tag{2.41}$$

Here, we see that the new terms for astigmatism arising from stop shift are the products of the coma system sum with entrance pupil shift, and of the spherical aberration sum with entrance pupil shift squared. The same result applies to field curvature, by inspection.

Finally, with distortion, we get:

$$\sum(x_i - \Delta_{EP})^3 W_i = \sum x_i^3 W_i - 3\Delta_{EP}\sum x_i^2 W_i + 3\Delta_{EP}\sum x_i W_i - \Delta_{EP}^3\sum W_i. \quad (2.42)$$

We see the pattern; the stop shift terms for any rank are linear with the stop shift in the next rank down, quadratic in the stop shift for the term two ranks down, and with distortion, we have a cubic term in the stop shift with the total spherical aberration sum, which is three ranks down.

Considering the case of the paraboloid mirror that we investigated above, we now see that because the system sum for spherical aberration was zero, moving the stop made no change to the system coma. However, because the coma system sum was nonzero, we could change astigmatism by moving the stop, and we could find a stop location that cancelled astigmatism.

We can further infer that if we were to put a coma corrector lens on the paraboloid mirror, so that the system sums for coma and spherical aberration are both zero, the astigmatism would be independent of the stop position.

When we have no spherical aberration, coma, or astigmatism in the system, we can put the stop anywhere we want, and none of the third-order aberrations are affected.

This fundamental property of optical systems is very useful in plate diagram analysis and design, as the stop can be placed on mirror vertices or centres of curvature, thereby eliminating contributions to system sum equations and thus reducing their complexity. Examples of this will occur naturally throughout this book.

2.7 Summary

In this chapter, a complete toolset has been developed from first principles that allows the construction and analysis of plate diagrams for multi-element systems. In the following chapters of this book, these tools will be used for this purpose. The presentation here was deliberately careful and avoided skipping detail, to the point of pedantry. This was a deliberate choice, as it was considered that it is less harmful for an expert to be unhappy about being given too much detail than it is for a student to fail to grasp the method for lack thereof.

Going forward, having done the hard work of derivation, the pace and content will hopefully be of interest to people at a wide range of levels.

The particular equations from this chapter that will be used most regularly are:

- Equation (2.13) for matrix ray tracing.
- Equation (2.9) is also useful for solving for foci ($y = 0$).
- Equation (2.23) gives angles of incidence on mirrors in convergent or divergent light.

- Equations (2.28) and (2.29) for spherical aberration in convergent/divergent and collimated light, respectively.
- Equation (2.32) for off-axis aberration.
- Equation (2.33) for Petzval curvature.

References

[1] Gauss C F 1840 *Dioptrische Untersuchungen Carl Friedrich Gauss: Werke* vol 5 (Göttingen: Königliche Gesellschaft der Wissenschaften) 1–30

[2] Seidel L 1857 von Über die Theorie der Fehler, mit welchen die durch optische Instrumente gesehenen Bilder behaftet sind, und über die mathematischen Bedingungen ihrer Aufhebung Abhandlungen der Mathematisch-Physikalischen Classe der Königlich Bayerischen Akademie der Wissenschaften **4** 227–67

[3] Conrady A E 1929 *Applied Optics and Optical Design* (Oxford: Clarendon)

[4] Kingslake R and Johnson R B 2010 *Lens Design Fundamentals* 2nd edn (Burlington, MA and Oxford: Academic Press/SPIE)

[5] Burch C R 1979 Application of the plate diagram to reflecting telescope design *Opt. Acta* **26** 493–504

[6] Schmidt B 1931 Ein lichtstarkes komafreies *Spiegel System Mitt. Hamb. Sternw.* **7** 15–21

[7] Linfoot E H 1955 *Recent Advances in Optics* (Oxford: Clarendon)

[8] Hopkins H H 1950 *Wavefront Aberration Coefficients* (Oxford: Clarendon)

[9] Rogers J R and Rakich A 2020 The importance of Petzval correction in generalized Offner designs *Proc. SPIE 11482 Current Developments in Lens Design and Optical Engineering XXI*

[10] Kingslake R 1874 *A History of the Photographic Lens* (Academic Press, 1989, p 17) (footnote cites *Brit. J. Phot. Almanac* p 44)

[11] Burch C R 1943 On aspheric anastigmatic systems *Proc. Phys. Soc.* **55** 433–44

[12] Lemaître G R 2009 *Astronomical Optics and Elasticity Theory: Active Optics Methods Astronomy and Astrophysics Library* (Berlin, Heidelberg: Springer) 2

[13] Schwarzschild K 1905 Untersuchungen zur geometrischen Optik I. Systematische Entwicklung der Abbildungslehre im Anschluss an die konforme *Abbildung Math. Ann.* **60** 561–605

[14] Rakich A 2005 The 100th birthday of the conic constant and Schwarzschild's revolutionary papers in optics *Proc. SPIE* **5875** 587501

[15] Aldis H L 1900 On the construction of photographic objectives *Photogr. J.* **40** (vol. 24 new series) 291–9 https://archive.rps.org/archive/volume-40?

Chapter 3

Multi-element telescope systems

In this chapter, the tools developed so far will be used to investigate a wide range of reflecting telescope optics. It is hoped that through this chapter, readers who have not previously used the plate diagram will come to appreciate its versatility as an analytical tool and gain new insights into the fundamental properties of optical systems. Worked examples are provided, and hopefully, the reader will be motivated to try their own examples. The intent of this chapter is not so much to present an encyclopaedic discourse about reflecting telescopes but rather to show the inner workings of reflecting systems from the perspective of the plate diagram.

3.1 Classical two-mirror telescopes

We will start by looking at the plate diagrams for two classical telescopes: the Cassegrain and the Gregorian. These telescopes both have a concave paraboloidal primary mirror, which, as we know, is represented by two plates in the plate diagram. Additionally, they both have aspheric secondary mirrors.

Both mirrors work by making one geometric perfect conjugate of each conicoid mirror coincide. The paraboloid mirror in both cases has one perfect focus at infinity. It produces an image that is perfect on axis at half its radius. The Cassegrain telescope has a convex hyperboloidal secondary mirror (conic constant, $k < -1$), which is placed before the focus of the primary mirror and produces an image through a hole in the primary mirror. The virtual near focus of the convex hyperboloid is coincident with the real focus of the paraboloidal primary mirror.

All rays of the axial pencil converging towards the focus from the primary mirror are thus aimed at the hyperboloid's near focus. When they reflect from this mirror, they form a perfect image at the real, far focus of the hyperboloid.

The Gregorian telescope has a concave ellipsoidal mirror ($-1 < k < 0$). A Gregorian has a similar mode of operation, but this time, the secondary mirror is concave and after the prime focus, so both the near focus and the far focus of the conicoid are real, finite, and separated, making it an ellipsoid.

doi:10.1088/978-0-7503-3099-2ch3

For this example, two systems have been set up in ray-tracing software. They produce a final focus with a focal ratio of f/10 and have f/2 primary mirrors 1 m in diameter. Figure 3.1 shows the layout diagrams, figure 3.2 shows the spot diagrams, and figure 3.3 shows the aberrations contributed by the primary mirror and secondary mirror and the system sum. Various points are discussed in the captions.

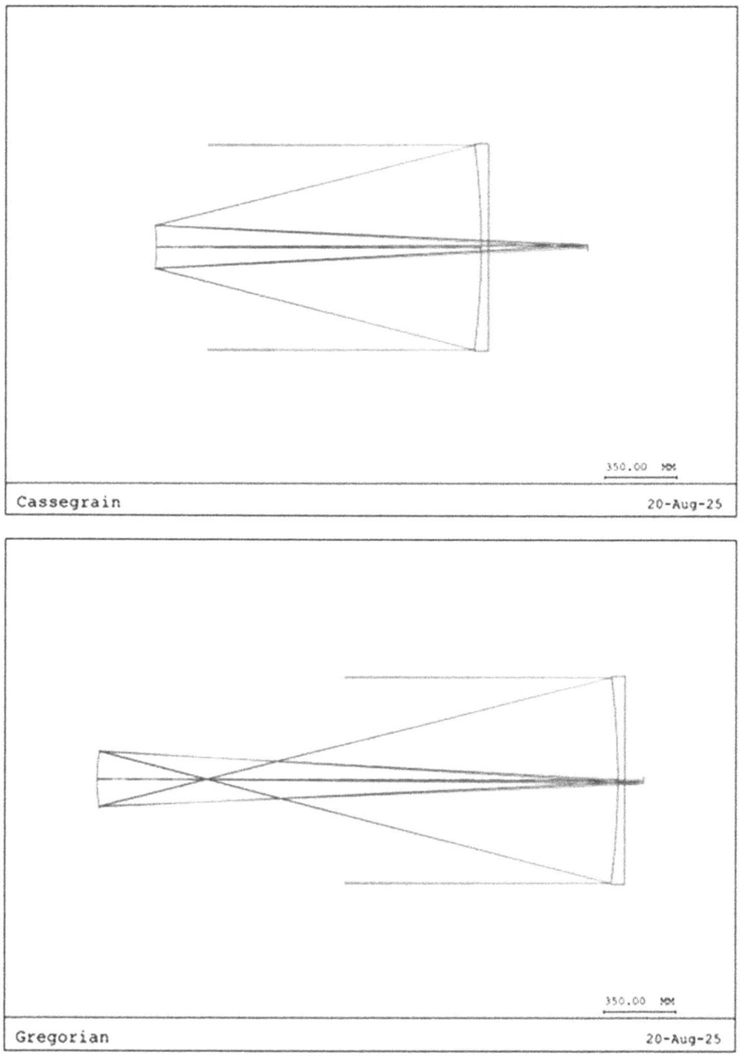

Figure 3.1. Optical layouts, roughly to the same scale, for f/10 telescopes. Both telescopes have a concave paraboloidal primary mirror. The Cassegrain telescope (top) has a hyperboloid secondary mirror, with its near (virtual) focus laid exactly on top of the focus of the paraboloid mirror, so it redirects light stigmatically to its second focus. The Gregorian telescope does the same thing, but this time the concave ellipsoid secondary mirror has two real foci, and the near focus is located at the primary mirror focus. Both telescopes produce on-axis stigmatic images via the well-known properties of conic sections. The paraboloidal primary mirror has its object conjugate at infinity. Created with Code V.

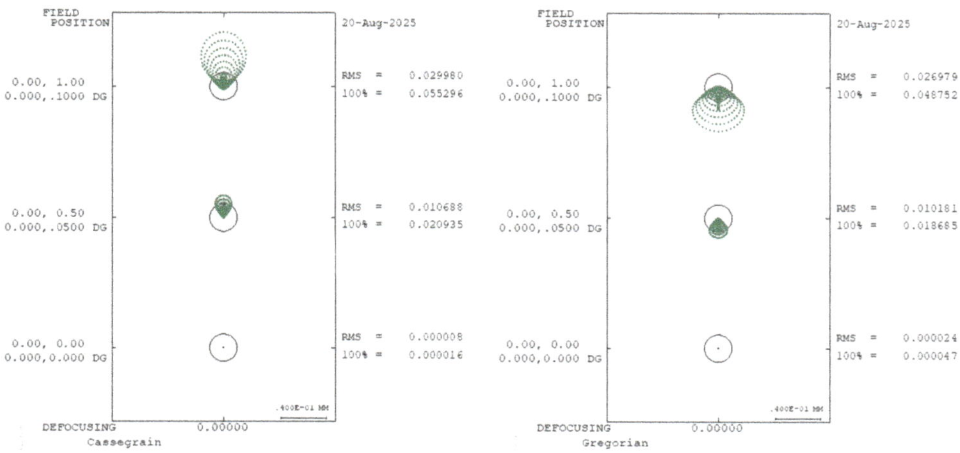

Figure 3.2. Spot diagrams for a 0.1 degree field for the Cassegrain (left) and Gregorian (right). Both have similar amounts of field aberration, with coma dominating. Created with Code V.

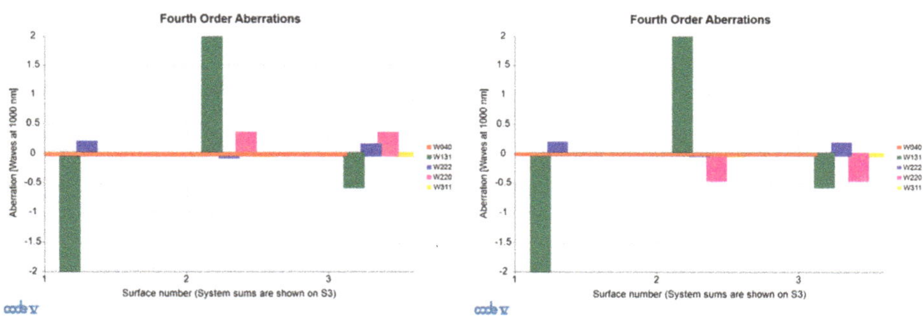

Figure 3.3. Seidel diagrams for both telescopes, Cassegrain (left) and Gregorian (right). The diagrams have been magnified so that the sums in the image (right of each diagram) are not swamped by the individual contributions from M1 (left) and M2 (middle). Created with Code V.

Considering figure 3.3, we can see that each of the mirrors is individually corrected for spherical aberration. In both telescopes, large amounts of coma from each mirror cancel, leaving a small residual of the same sign in their respective images. Conversely, only the primary mirrors contribute any astigmatism. Field curvature switches sign and is the only aberration that is noticeably worse in the Gregorian than in the Cassegrain. However, as the field curvature and astigmatism partially cancel each other in the Gregorian and sum in the Cassegrain, overall, the image blurs are comparable.

Table 3.1 lists the optical parameters for the two telescopes, as required to construct the plate diagram.

We shall now produce a plate diagram for the Cassegrain telescope. We already have all the information we need to make the plate diagram for the paraboloid mirror, as we did in the previous chapter. The required steps are:

Table 3.1. Parameters for the Cassegrain and Gregorian telescopes under consideration.

Parameter	Cassegrain	Gregorian
M1 radius	−4000 mm	−4000 mm
M1 semi-diameter	500 mm	500 mm
M1 conic constant	−1	−1
Stop location	At M1	At M1
M1–M2 separation	−1579.737 17 mm	−2532.332 92 mm
M2 radius	−1050.821 47 mm	887.129 00 mm
M2 conic constant	−2.251 17	−0.444 21
Field of view	0.1 degrees $\left(\frac{\pi}{1800}\text{ radians}\right)$	0.1 degrees $\left(\frac{\pi}{1800}\text{ radians}\right)$

1) Calculate W_{S1} and W_{A1}, spherical aberration contributions from the primary mirror base sphere and aspheric profile, respectively.
2) Assign correct values for x_{S1} and x_{A1}, the corresponding plate distances from the entrance pupil.
3) Calculate the paraxial quantities i_2 and y_2 at M2.
4) Using these, calculate W_{S2} and W_{A2}, spherical aberration contributions from the secondary mirror base sphere and aspheric profile, respectively.
5) Calculate x_{S2} and x_{A2}, images of the centre of curvature of M2 and the vertex of M2, respectively, in object space (that is, imaged through the primary mirror).

With these five steps complete, we will be able to produce the plate diagram for the system or, as may be the case here, do something interesting with the analysis.

We shall proceed step by step. Note that in this book, we follow the convention that equations in worked examples are left unnumbered; only equations that define concepts or relationships are numbered.

Step (1)

$$W_{S1} = -\frac{1}{4} \times 1 \times \frac{1}{-4000^3} \times 500^4 = 0.244141 \text{ mm}.$$

We can take a shortcut here because, in this case, we know that $W_{S1} = -W_{A1}$,

$$W_{A1} = -0.244141 \text{ mm}.$$

Step (2)

The aperture stop is on the primary mirror, so, in this case, it is also the entrance pupil. The plate for the base sphere is at the centre of curvature of the primary mirror and the plate for the asphericity is at the primary mirror:

$$x_{S1} = -4000 \text{ mm},$$

$x_{A1} = 0,$ because the stop is at the surface.

Step (3)

In this step, we can use the matrix approach from equation (2.13) and the relation $i = \varphi - u$ from equation (2.23). We start with collimated light, so $u = 0$, and marginal axial ray at height $y_c = 500$ mm. The refractive index in our initial space is 1.

$$(0,\ 500).\begin{pmatrix} 1 & 0 \\ \frac{-2}{-4000} & 1 \end{pmatrix}\begin{pmatrix} 1 & -\frac{-1579.737\,17}{-1} \\ 0 & 1 \end{pmatrix} = (0.25,\quad 105.066).$$

Here, the first matrix is the power matrix from M1, and the second matrix is the transfer matrix, using the M1–M2 separation from table 3.1. A trick that catches many people here is we now have a quantity of 0.25, but that is NOT the angle u', rather it is $n_2 u'$, and $n_2 = -1$ here. So $u' = -0.25$. Considering the layout diagram in figure 3.1, we can see that this is the sign that we should expect for u_2.

Proceeding, we use the relation $i = \varphi - u$ to calculate i, noting that $\varphi = \frac{y}{r}$:

$$i_2 = \frac{105.065\,70}{-1050.821\,47} - (-0.25) = 0.150\,02.$$

Again, a sanity check shows that the ray, travelling from right to left, should be above the surface normal to the right of the surface and below it to the left, so i_2 is positive according to the sign convention shown in figure 2.5.

We have now $i_2 = 0.150\,02$ radians and $y_2 = 105.066$ mm.

Step (4)

$$\begin{aligned} W_{S2} &= -\tfrac{1}{4} n_2 c_2 i_2^2 y_2^2 \\ &= -\tfrac{1}{4} \times -1 \times \frac{1}{-1050.821\,47} \times 0.150\,02^2 \times 105.065\,70^2 \\ &= -\ 0.059\,106 \text{ mm.} \\ W_{A2} &= k \times -\tfrac{1}{4} n_2 c_2^3 y_2^4 \\ &= -\ 2.251\,17 \times -\tfrac{1}{4} \times -1 \times \frac{1}{-1050.821\,47^3} \times 105.065\,70^4 \\ &= = 0.059\,102 \text{ mm, which is close enough...} \end{aligned}$$

Step (5)

In this step, we could use Newton's lens formula, but it is advisable to keep practising with the matrix ray tracing. We calculate x_{S2}, the image of the centre of curvature of M2, through M1 into object space. First, we calculate u' and y, the values after reflection at M1, then we can rearrange the transfer equation $y_+ = y - u't'$, and solve for $y_+ = 0$. We pick an arbitrary nonzero value for u_1 and start at $y_1 = 0$. The distance from the centre of curvature of M2–M1 is the mirror separation plus M2's radius of curvature. For our raytrace, we first produce a transfer matrix to transfer the ray to M1, then we produce the power matrix, to calculate values after reflection at M1.

When producing our transfer matrix, we take careful note to reverse the sign from the values given in table 3.1, because we are going FROM the centre of curvature of M2 TO the surface of M1, in this case. As the launch ray is travelling from left to right initially, our first refractive index is $n_1 = +1$.

$$(-0.1,\ 0)\begin{pmatrix} 1 & -\frac{-(1579.737\,17 + 1050.821\,47)}{1} \\ 0 & 1 \end{pmatrix}\begin{pmatrix} 1 & 0 \\ \frac{-2}{-4000} & 1 \end{pmatrix} = (0.031\,528,\quad 263.056).$$

Remembering to divide the angle result by the refractive index, -1, we get from this result that

$$x_{S2} = \frac{263.056}{-0.031\,528} = -8343.58\ \text{mm}.$$

Note that this gives us the distance from M1 where y is zero in object space, so this is the axial location of the image of the centre of curvature of M2.

Step (6)

We can do a similar trace to find the image of the surface vertex of M2 in object space, giving us x_{A2}:

$$\left(-0.1,\ 0\right)\begin{pmatrix} 1 & -\frac{(1579.73717)}{1} \\ 0 & 1 \end{pmatrix}\begin{pmatrix} 1 & 0 \\ \frac{-2}{-4000} & 1 \end{pmatrix} = \left(-0.02101,\ 157.9737\}\right).$$

$$x_{A2} = \frac{157.9737}{0.02101} = 7517.85\ \text{mm}.$$

Table 3.2 immediately below collects the relevant results. At this point we can calculate the aberrations of the system.

These system-sum values can be compared with the output from ray-tracing software given in table 3.3.

Any slight differences that arise are due to rounding error propagation. When such calculations are being done in earnest, instead of being copied and pasted through various programs as happens when producing a book, full numerical accuracy can be preserved, and the match is exact. Note that the sign difference for the coma term simply reflects the fact that the ray-tracing software was baselined using a different sign for the maximum field angle than that used in this calculation. In the case of the astigmatism term, the field angle is squared, so the sign difference vanishes.

Table 3.2. The table compiles quantities calculated in steps 1–6 and lists system sums for aberration.

Quantity	M1 S	M1 A	M2 S	M2 A
W (mm)	0.244 141	−0.244 141	−0.059 106	0.059 102
Spherical aberration				
$\sum W$	0.0000 mm			
x (mm)	−4000	0	−8343.58	7517.85
Coma				
Wx	−976.564	0	493.155 639	444.319 97
$-4\frac{u}{y_c}\sum Wx$	0.000 545 777 mm			
Astigmatism				
Wx^2	3906 256	0	−4114 683.5	3340 330.9
$4(\frac{u}{y_c})^2\sum Wx^2$	0.000 152 645 mm			

Table 3.3. Ray-tracing software results for system sums. Note that the software displays these aberration coefficients in units of waves, so for display purposes, the wavelength is set to 1 micron, which is easily scaled by eye to match the equations. In the table, W040 is spherical aberration, W131 is coma, W222 is the astigmatism we calculate (W220 is the field-curvature wavefront coefficient), and W311 is distortion.

THIRD ORDER IMAGE ABERRATIONS (Wavelengths at 1000 nm)					
Surface	W040	W131	W220	W222	W311
SUM	-0.0001	-0.5461	0.3435	0.1527	--0.0110

As noted, the individual mirrors are currently corrected for spherical aberration, with aspheric W cancelling that of the base spheres, at each mirror.

Considering the coma sum, we see that the sum of Wx is nonzero:

$$Wx = -39.0884 \text{ mm}^2.$$

To the first order, all mirrors are spherical. The aspheric terms only begin to perturb wavefronts at the third order of the expansion of $\sin\theta$. Therefore, we are free to manipulate the conic constants of both M1 and M2 to try to achieve a better result without disturbing the layout of the system.

As an interesting example of this, we can consider another form of two-mirror telescope that has a significant design simplification: the Dall–Kirkham. This telescope has a spherical secondary mirror, which is considerably simpler to manufacture and align than the aspheric secondary mirror required by any other two-mirror telescope in this family. This telescope does NOT work via the conjugation of perfect conicoid foci, as the secondary mirror is spherical.

Still, we can see how it works without doing any more maths but simply by looking at the plate diagram tabulated in table 3.2. With this information, we can consider how things change when we make M2 spherical without changing radii or the air space of the system (thus maintaining the first-order properties unchanged). W_{A2} is to be made zero in this case; therefore, to maintain the balance, the spherical aberration W_{A1} has to be reduced, which is achieved by making M1 ellipsoidal ($-1 < k < 0$). With the stop on M1 ($x_{A1} = 0$), any change in W_{A1} has no impact on field aberrations, but the zeroing of W_{A2} removes the compensating contributions of both coma and astigmatism from that column in table 3.2. Compared to the classical Cassegrain, coma now increases by a factor of approximately ten, and astigmatism also degrades a little. This is immediately visible just by looking at these plate parameters. While most telescope designers are familiar with the Dall–Kirkham, this separation of the aspheric and spherical contributions to aberration, which is natural with the plate diagram, does not occur naturally with ray-tracing analysis, so this may provide new insight even for experienced designers.

This sort of 'by-eye' analysis is uniquely made possible by the plate diagram. The author is unaware of any other analytical technique by which such information is made so immediately visible. The output from ray tracing software, as well, always conflates the aberrations arising at an optical surface from the spherical and aspherical components into one result, making the sort of insight achieved here, into the contributions of each component, impossible.

The two-mirror telescope corresponds to four plates, and the plate equations can be reduced to three equations in three unknowns, with suitable choices in base geometry. It is therefore possible to simultaneously correct three aberrations, and this will be a main topic of subsequent sections of this book. First, though, we will consider what is required to simultaneously correct only two aberrations: spherical aberration and coma.

3.2 Aplanatic two-mirror telescopes

The simultaneous correction of spherical aberration and coma is referred to as 'aplanatic correction', and systems with such correction are known as 'aplanats'. Any spherical optical surface is aplanatic for one particular pair of object/image conjugate points [1]. The aplanatic modification of the classical Cassegrain telescope was first found analytically by Chrétien in 1910 [2], who investigated the possibility of producing an aplanatic telescope at the suggestion of his friend, Ritchey, and has been named the 'Ritchey–Chrétien' telescope [3]. Schwarzschild had actually produced all the necessary formulas in 1905 [4] but had never considered this particular solution, being, at the time, more interested in anastigmatic systems. The Ritchey–Chrétien telescope went on to become the main workhorse of 20th-century astronomy, offering significant improvements in image quality in the near field compared to the classical form.

We can quickly arrive at a Ritchey–Chrétien solution using the plate diagram. Considering $W_i x_i$ in table 3.3, we can note it is clear that M1 A makes no contribution to off-axis aberration. This is expected when we consider that, because the system stop is at M1, x_{A1} is zero. Keeping the vertex spherical radii of the mirrors fixed (fixed first-order properties) means that we can only adjust coma by adjusting W_{A2}, which we achieve by adjusting the conic constant of M2.

In doing so, we unbalance the spherical aberration sum, but this can be driven back to zero by adjusting W_{A1}, i.e. by altering its conic constant. Rebalancing the spherical aberration in this way does not disturb coma at this stage, because $x_{A1} = 0$.

To calculate the k_{A2new} required to drive coma to zero, we note that the initial value of k_{A2} is a factor in W_{A2}:

$$W_{A2} = k_{A2} \times -\frac{1}{4} n_2 c_2^3 y_2^4.$$

We also can note that

$$x_{A2} W_{A2} = 444.31997,$$

and finally, that the value of $\sum x_i W_i = -39.0884$.

From this we can see that we want to find a new M2 conic constant, k_{A2new}, such that

$$x_{A2} W_{A2new} = 444.31997 + 39.0884, \quad \text{so that } \sum x_i W_i \to 0.$$

Given that we are only changing the conic constant, we can easily find the new value of k_{A2} as follows:

$$k_{A2\,\mathrm{new}} = \left(1 + \frac{39.0884}{444.319\,97}\right) k_{A2} = -2.449\,213.$$

To check the new value, we can recalculate W_{A2}:

$$\begin{aligned}
W_{A2\,\mathrm{new}} &= k_{A2\,\mathrm{new}} \times -\tfrac{1}{4} n_2 c_2^3 y_2^4 \\
&= -2.449\,213 - \tfrac{1}{4} \times -1 \times \frac{1}{-1050.821\,47^3} \times 105.065\,70^4 \\
&= 0.064\,301\,9, \text{ from which} \\
W_{A2\,\mathrm{new}} x_{A2} &= 0.064\,301\,9 \times 7517.85 \\
&= 483.411\,79.
\end{aligned}$$

When we sum this value with the two corresponding values from M1 in table 3.2, we get

$$\begin{aligned}
\sum Wx &= -976.564 + 493.155639 + 483.41179 \\
&= 0.003429.
\end{aligned}$$

The residual has been reduced by a factor of 10 000 which is approximately our rounding error, and we can conclude that coma is now zero.

The spherical aberration sum now becomes

$$\begin{aligned}
\sum W &= 0.244\,141 - 0.244\,141 - 0.059\,106 + 0.064\,301\,9 \\
&= 0.005\,195\,9.
\end{aligned}$$

We can now adjust the conic constant of M1 from −1 (paraboloid), as we did previously for k_{A2}, to rebalance spherical aberration:

$$k_{A1\text{ new}} = \left(1 + \frac{-0.005\,195\,9}{-0.244\,141}\right)k_{A1} = -1.021\,282.$$

To check the new value, we can calculate $W_{A1\text{ new}}$:

$$\begin{aligned} W_{A1\text{ new}} &= \Delta k_{A1\text{ new}} \times -\tfrac{1}{4}n_1 c_1^3 y_1^4 \\ &\quad - 1.021\,282 \times 0.25 \times 1 \times \tfrac{1}{-4000^3} \times 500^4 \\ &= -\,0.249\,337, \end{aligned}$$

from which we find that

$$\begin{aligned} \textstyle\sum W &= 0.244\,141 - 0.249\,337 - 0.059\,106 + 0.064\,3019 \\ &= 0. \end{aligned}$$

Entering the two new conic constants in ray-tracing software yields the results illustrated in table 3.4 and figures 3.4 and 3.5.

A Gregorian aplanat can be derived by exactly the same method. This is left as an exercise for the reader. It can be done more simply. The worked example given here has been deliberately pedantic, with the goal of enabling students of this topic to follow each step exactly and not get stuck at any point, while also becoming familiar in detail with various concepts involved at the 'nuts and bolts' level. In fact, the solution for the aplanatizing conicoids can be found in two steps. To demonstrate, once we have calculated the quantities in table 3.2 and summed the Wx product row, obtaining 39.0884, we can go straight to

$$k_{A2\text{ new}} = \left(1 + \frac{39.0884}{444.319\,97}\right)k_{A2} = -2.449\,213,$$

and

$$k_{A1\text{ new}} = \left(1 + \frac{W_{A2}\left(1 - \frac{k_{A2\text{ new}}}{k_{A2}}\right)}{-0.244\,141}\right)k_{A1} = -1.021\,282,$$

Table 3.4. Comparing these values to those in table 3.2, we see that spherical aberration and coma now both have values dominated by the rounding errors that have arisen from 'cut-and-paste' display equations; however, they are still very small values. The astigmatism value has increased by a few percent compared to the original system.

THIRD ORDER IMAGE ABERRATIONS (Wavelengths at 1000 nm)					
Surface	W040	W131	W220	W222	W311
SUM	0.0000	-0.0001	0.3507	0.1670	-0.0106

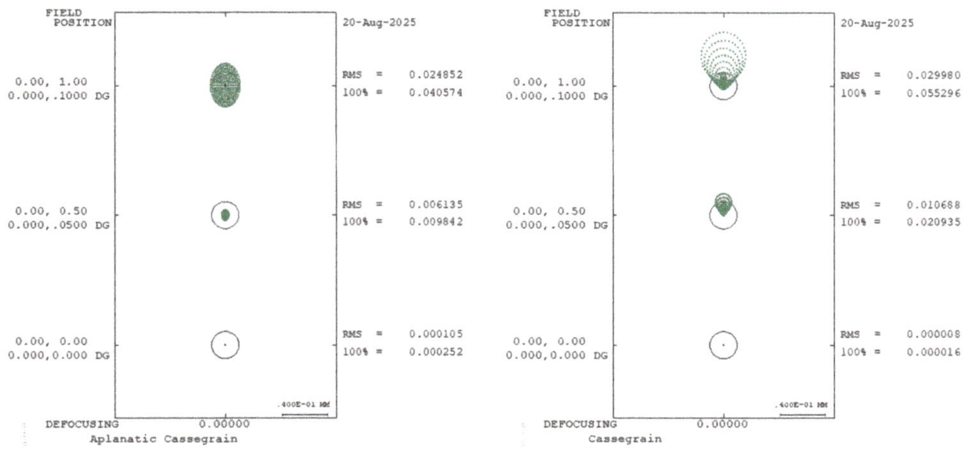

Figure 3.4. New aplanatic Cassegrain (left) compared to the original classical Cassegrain (right). Coma is now removed, leaving a residual of astigmatism and field curvature. Created with Code V.

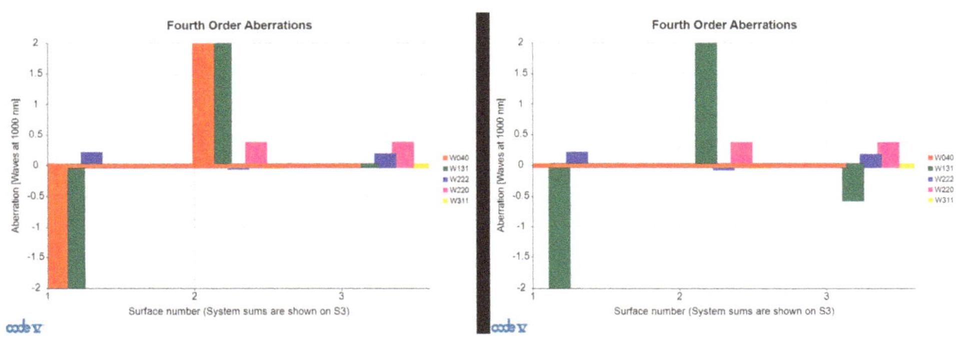

Figure 3.5. The Seidel diagram for the aplanatic Cassegrain (left) compared to the original (right) shows that both M1 and M2 now have 'off-the-chart' spherical aberration, but that this is balanced in the final image. Coma (green) is now gone from the sum on the left. Created with Code V.

which, as we have now seen, is a verified correct result. To produce the Gregorian aplanat solution, Steps 1–5 above can be repeated using the values given in table 3.1, and the resultant values can be used to produce an equivalent table to table 3.2. From there, Steps 1 and 2 directly above yield the desired new conic constants.

It can be observed that if we are to keep this fixed geometry with the radii and air space locked, it is possible to correct *either* coma *or* astigmatism (following the exact same method as we used for coma, but now with the Wx^2 sum as the starting point), but not both at once.

Coma has a linear dependence on field angle, whereas astigmatism and field curvature have a quadratic dependence. This means that coma is the dominant aberration in the near-axis field points and astigmatism and field curvature overtake coma at some radial zone in the field, dominating the outer field. For this reason, aplanats, and then anastigmats, are of interest for increasingly wide-field systems, but a system corrected for only spherical aberration and astigmatism/field curvature is not practically interesting for imaging applications and has not been named. Some authors refer to an 'aplanatic anastigmat', implying that both names are required to describe the simultaneous correction of coma and astigmatism (together with spherical aberration), but the common convention is that 'anastigmat' alone refers to a system corrected for all three aberrations. These systems are considered next.

3.3 Anastigmatic systems

We have seen that in the case of a Schmidt telescope, we can simultaneously correct spherical aberration, coma, and astigmatism with only two plates. The Schmidt plate is coincident with the plate representing the mirror, and the two exactly cancel. We have also seen that when these plates are not coincident, as is the case for a paraboloid mirror, it is impossible to simultaneously correct coma and astigmatism (though it is possible to correct astigmatism by placing the aperture stop midway between the two equal-power plates).

Also, in the last section, we saw how we could control coma and spherical aberration, or coma and astigmatism, with two plates, but not all three of these aberrations simultaneously. A fascinating result, discovered by Aldis [5] (or possibly Petzval) and discussed by Burch [6], has not, to the author's knowledge, been explicitly proven using any other analytical approach. Burch calls it the 'Aldis 4-Plate Theorem'. If we consider the first three aberrations arising from a single plate, as expressed in equation (2.29):

$$\frac{W}{y_c^4}(\rho^4) \rightarrow \frac{W}{y_c^4}\left(y_c^4\right) = W \quad \text{spherical aberration}$$

$$\frac{W}{y_c^4}(4y_p\rho^3\mathrm{Cos}^3\theta) \rightarrow -4\frac{W}{y_c^4}ux\left(y_c^3\right) = -4\frac{u}{y_c}xW \quad \text{coma}$$

$$\frac{W}{y_c^4}\left(4y_p^2\rho^2\mathrm{Cos}^2\theta\right) \rightarrow 4\frac{W}{y_c^4}(ux)^2\left(y_c^2\right) = 4\left(\frac{u}{y_c}\right)^2 x^2W \quad \text{astigmatism}$$

and recall that x is the distance between the plate and the entrance pupil in the object space of a telescope, we can further think of W as being a force acting in a plane perpendicular to the x direction. If we consider the system sum of aberrations in a system of n plates being driven to zero and disregard the constant terms, we can express this as

$$\sum_{i=1}^{n} W_i = 0$$

$$\sum_{i=1}^{n} x_i W_i = 0$$

$$\sum_{i=1}^{n} x_i^2 W_i = 0.$$

We can see that the problem of producing an anastigmat from a system of non-coincident plates is exactly analogous to that of simultaneously zeroing the zeroth, first, and second moments of a system of non-coincident parallel forces, as shown in figure 3.6.

Figure 3.6(A) represents a system of parallel forces acting on a line. Each force F_i is exactly analogous to our W_i from the plate diagram. The forces are separated by distances a, b, and c as indicated.

If we consider the subset of (A) with three parallel forces acting, as shown in (B) and assign forces

$$F_1 = b, \quad F_2 = -(a + b) \text{ and } F_3 = a,$$

we can see that

$$\text{zeroth moment: } b - (a + b) + a = 0,$$

$$\text{first moment: } -ab + ba = 0,$$

and the second moment can be given by:

$$a^2 b + b^2 a = ab(a + b).$$

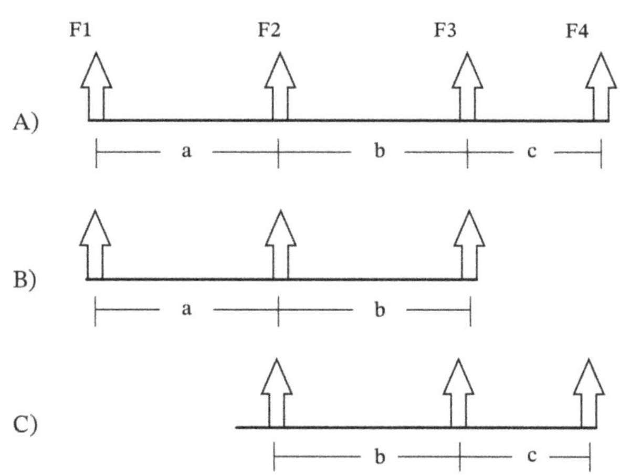

Figure 3.6. It is a long-standing result that four separate forces are required to drive the zeroth, first, and second moments of a system of parallel forces acting on a line to zero [7]. It is shown below how forces must be allocated such that this result is achieved.

The second moment can therefore only be made zero if $a = -b$, putting opposing, equal-magnitude forces on top of each other, or if $a = b = 0$, which is not interesting. The second moment can be made unity by dividing each force by $ab(a + b)$, giving:

$$F_1 = \frac{1}{a(a+b)},$$

$$F_2 = -\frac{1}{ab},$$

$$F_3 = \frac{1}{b(a+b)}. \tag{3.1}$$

Similarly, for the subset of (A) shown in diagram (C), if we assign forces $F_2 = -c$, $F_3 = (b + c)$ and $F_3 = -b$, we can see that the zeroth and first moments are again zero and the second moment can be written as

$$\text{second moment} = -b^2c - c^2b = -bc(b+c).$$

This moment can be made negative unity by dividing each force by $bc(b + c)$, giving

$$F_2 = -\frac{1}{b(b+c)},$$

$$F_3 = \frac{1}{bc},$$

$$F_4 = -\frac{1}{c(b+c)}. \tag{3.2}$$

We can consider the systems now as optical plate diagrams, and we see that each of the systems with the plate strengths allocated above is corrected for spherical aberration and coma; moreover, the system in (B) has astigmatism at unity, and the system in (C) has astigmatism at negative unity.

Combining these systems back into the original configuration in figure 3.6(A), the resultant system has the following forces:

$$F_1 = \frac{1}{a(a+b)}$$

$$F_2 = -\frac{1}{ab} - \frac{1}{b(b+c)},$$

$$F_3 = \frac{1}{bc} + \frac{1}{b(a+b)},$$

$$F_4 = -\frac{1}{c(b+c)}. \tag{3.3}$$

In this system, the zeroth and first moments from each subsystem remain zero when the systems are joined, and now the second moments cancel. This argument is sufficient to prove that for non-coincident plates of nonzero magnitude, a minimum of four plates is required to produce an anastigmat. We can also conclude that the strength of each plate is inversely proportional to the product of the distances between its position and those of all the other plates *in the aplanatising subsystems B and C*. This last statement is emphasised because it is misstated by Burch (see reference [6]), who otherwise gives the correct mathematics but states that each plate's strength is inversely proportional to the distance of *all the other plates*, which was a source of some confusion to the author at first.

A final conclusion that we can draw from this mechanical analogy is that the plate strengths must alternate in sign from left to right in a system of four non-coincident plates with zeroed moments.

To the author's knowledge, this important set of conclusions has not been reached via any other analytical pathway. The consequences for optical design are significant. We can immediately conclude that the set of 'simplest possible anastigmats', discounting systems with coincident plates, is limited to systems with the following configurations.

3.3.1 Catoptric systems

Two aspheric mirrors.
Three mirrors, one of which is a conicoid.
Four spherical mirrors.

3.3.2 Catadioptric systems with one or more Schmidt plates

Two mirrors, one aspheric, and one real Schmidt plate.
Two spherical mirrors and two Schmidt plates.
Three spherical mirrors and one Schmidt plate.

3.3.3 Dioptric systems without Schmidt plates

Two lens elements with spherical surfaces.
One lens element with two aspheric surfaces.

3.3.4 Other combinations

One aspheric mirror, one spherical lens element.
One spherical mirror, one lens element with one aspheric surface
Two spherical mirrors, one spherical lens element…

More combinations are possible, but it is a finite set. The interesting point about this finding is that there is very little literature regarding some of these combinations. The catoptric systems have all been well described, and the mirror/plate combinations have been covered in detail by Linfoot, but in a relatively difficult-to-access version of the plate-diagram formalism [8].

The singlet anastigmat lens has also been described, first by Burch, but it would still come as a surprise to many optical designers. The plate diagram finds its greatest strength in systems without powered refracting components, as there is no chromatic aberration to contend with. Still, given the relative simplicity of the method, including achromatising constraints in a plate-diagram-based system does not pose too much difficulty, as Burch shows (see [6]).

Even for refracting or catadioptric monochromats, the range of use cases continues to expand as lasers continue to find more applications. If this book runs to subsequent editions, a development of refracting-system plate diagrams will be included.

The rest of this chapter will explore the first type of system: reflecting anastigmats. The main goals of this book are to make the plate diagram accessible to interested readers and to demonstrate its usefulness in building an intuitive understanding of how aberrations arise in optical systems; for these purposes, the reflecting anastigmats are a good type of system to gain familiarity with.

3.4 The Mersenne confocal paraboloidal pair (beam expander)

While most people familiar with telescopes are aware of the names ‘Cassegrain’ and ‘Gregory’ as the inventors of the two eponymous two-mirror telescope forms, fewer people have heard about Marin Mersenne’s efforts in this direction. Mersenne (1588–1648) was a contemporary and friend of René Descartes. He is famous for his work in prime number theory; ‘Mersenne primes’ are named after him [9]. In ‘La Dioptrique’, Descartes had applied his newly invented analytical geometry to show the geometric form required of lens surfaces to produce images free of spherical aberration: the Cartesian oval [10]. Mersenne immediately saw that the same approach could be applied to systems of mirrors, and in 1636 published two versions of confocal paraboloidal pairs [11] in what later came to be known as the Gregorian (1663) [12] and Cassegrain (1672) [13] forms when they are produced in focal mode.

Mersenne’s confocal paraboloid mirrors produce an afocal system: a collimated pencil of rays is received by a large concave paraboloid mirror M1 and redirected to either a smaller convex paraboloid mirror before focus or a smaller concave paraboloid mirror after focus. As the mirrors are confocal, the smaller mirror reconjugates the light to infinity, and as both mirrors are paraboloids working at their natural conjugates, there is no spherical aberration (figure 3.7).

What Mersenne did not know, and could not have known, was that in his afocal forms he had produced a system with significant advantages over the focal forms that were to come later, in that the afocal forms not only corrected spherical aberration but also coma and astigmatism. Thus, the world’s first two-mirror telescope designs were anastigmatic, which is remarkable, given that coma and astigmatism were not formally identified as independent aberrations until the early 1800s. In this sense, it is fair to say that Mersenne was not just a genius but also a lucky one!

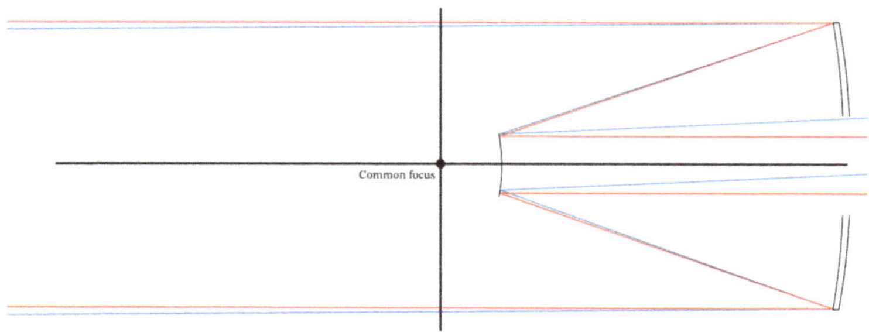

Figure 3.7. A Mersenne afocal anastigmat in 'Cassegrain' form, consisting of a concave primary paraboloid mirror confocal with a convex paraboloidal secondary mirror.

An excellent description of Mersenne's invention, and indeed a comprehensive history of reflecting telescopes in general, is given in Ray Wilson's seminal work *Reflecting Telescope Optics I*, which, while now 20 years old, remains the standard work in the field [2]. The author and Dr Wilson had long discussions on the topic of Mersenne's anastigmat, which Wilson regarded, along with the Schmidt telescope, as representing the fundamental basis for the aberration theory of all reflecting telescope systems.

The second edition of Wilson's book has a very interesting footnote on the 'historical discovery' of the Mersenne system [2]. Wilson identifies Baker as the first in the literature to mention the anastigmatic properties of confocal paraboloid mirror pairs, and Baker himself points to McCarthy at McDonald Observatory [14] as the first to recognise and use this property and to write about it in internal reports.

Given the work we have already done in producing plate diagrams for Cassegrain and Gregorian systems in section 3.1 of this chapter, it is a natural step now to consider the Mersenne designs, which are the limiting case of these designs, since the focal length goes to infinity.

We can use the plate diagram for a Mersenne system to produce a 'visual proof' of its anastigmatic properties. This proof requires one concept from first-order optics that was omitted from chapter 2: the 'Smith–Helmholtz–Lagrange' invariant, commonly referred to as the Lagrange invariant but independently discovered by all three [15]. For simplicity, we will refer to it as the Lagrange invariant in the following. Figure 3.8 gives quantities used to define the Lagrange invariant.

Using the sign conventions laid out in chapter 2 and taking the origin to lie at the axial point on Surface 1, we can see that

$$\begin{aligned} h1 &= (l_{ep1} - l_1)u_{p1} \\ &= \left(\frac{y_{p1}}{u_{p1}} - \frac{y_1}{u_1}\right)u_{p1} \\ &= \frac{u_1 y_{p1} - u_{p1} y_1}{u_1}. \end{aligned} \tag{3.4}$$

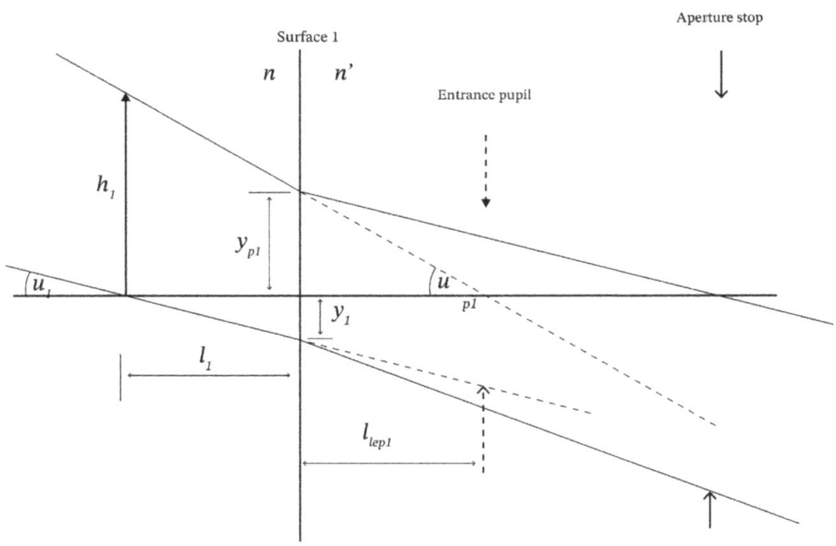

Figure 3.8. First-order quantities used to derive the Lagrange invariant. At the left of the diagram, two rays are traced from the object (h_1). The refracting surface, Surface 1, lies between the aperture stop and the object space. The entrance pupil is the image of the aperture stop, through Surface 1, in the object space. The lower ray, making a positive angle u_1 with the optical axis, is an axial marginal ray, as it originates on axis at the object, passes through the edge of the entrance pupil in object space, and then passes through the edge of the aperture stop in image space. The second ray, originating at the top of h_1, can be called the 'chief ray'; it is the principal ray, or 'gut ray' of the most oblique pencil of rays originating from the extreme edge of the object. This ray makes an angle of u_{p1} with the axis.

Multiplying this through by n_1u_1, we obtain

$$n_1u_1h_1 = n_1(u_1y_{p1} - u_{p1}y_1). \tag{3.5}$$

When $y_{p1} = 0$, we are at a pupil and y_1 is the pupil's semi-diameter. Equation (3.2) becomes

$$n_1u_1h_1 = -n_1u_{p1}y_1. \tag{3.6}$$

The quantities on both sides of equations (3.2) and (3.3) are the Lagrange invariant. By inspection, the results in equations (3.1) and (3.2) are general; they propagate through any number of subsequent refracting surfaces and finally to a system image surface. The Lagrange value is invariant in any space in the optical system and is defined in object space (in air) as the product of the diameter of the object and the angular size of the entrance pupil seen at the object, or alternatively, as the product of the diameter of the entrance pupil and the angular size of the object as seen at the pupil.

A practical consequence of this, pertaining to the Mersenne beam expander, is that the beam expansion ratio, which is the ratio of the diameter of the exit pupil to the entrance pupil, is inversely proportional to the ratio of exit and entrance chief ray angles. In other words, as the beam diameter is reduced by the Mersenne, so the field angle is increased in the same proportion.

Armed with this knowledge, we can now consider a limiting case of the confocal paraboloidal pair in which both the concave primary mirror, M1, and the convex secondary mirror, M2, have the same radius. As they are now confocal, they must lie exactly on top of each other. This is not a practical optical system, as it has a 100% obstruction, and its beam expansion ratio is one. However, it serves a purpose in this discussion.

Lying exactly on top of each other, these paraboloid mirrors exactly cancel each other.

Anyone who needs to be convinced of this cancellation can first consider the base spheres and apply equation (2.29) to the concave mirror, then apply equation (2.28) to the convex mirror, remembering to change the sign of n after reflection by M1. Alternatively, considering light travelling from right to left (so n is reversed in sign from when light travels from left to right) and first striking the convex mirror, applying (2.29) shows immediately that the spherical aberration has the same magnitude but the opposite sign to that of the concave mirror. Thus, the spheres cancel each other, and the parabolisation on each mirror therefore produces cancelling spherical aberration too.

In this case, we have exactly the plate diagram for the paraboloidal primary mirror shown in figure 2.18, with another plate diagram laid on top of it; the second has equal magnitude but the opposite sign of W's when compared to the first.

The pairs of plates, at the centre of curvature and at the surface vertices, respectively, are coincident and exactly cancel each other. The system of confocal equal-radius paraboloids is clearly anastigmatic.

Now consider what happens when we scale the secondary mirror in radius, making it smaller while keeping it confocal with the primary mirror. In this case, the two mirrors separate, and the height of y_2, the marginal axial ray at M2, scales directly with r_2, the radius of M2. We get a system as shown in figure 3.7. The beam expansion ratio has scaled in proportion to the ratio of radii, r_2: r_1.

The Lagrange invariant tells us directly that in this case, the chief ray angle scales as the inverse of this beam expansion ratio, so

$$r_2 : r_1 = y_2 : y_1 = u_{p1} : u_{p2}.$$

For the rest of this discussion, we can drop the subscript for u, which will mean the principal ray angle at M2 exclusively.

Also, we know that $W = -\frac{1}{4}nc^3y^4$ and that $W_{A2} = -W_{S2}$ for a paraboloid conjugated to infinity.

We know that as r_2 decreases so $c = \frac{1}{r}$ increases, and therefore c^3y^3 is invariant with respect to the scale of r_2.

We can conclude that W_{S2} and W_{A2} both decrease in direct proportion to both r_2 and y_2.

We recall from the plate-diagram analysis of a single paraboloid mirror in chapter 2 that setting the aperture stop exactly between the centre of curvature and the surface vertex of the mirror cancelled the astigmatism of a single paraboloid. Therefore, as the two paraboloids of the Mersenne system are confocal, setting the aperture stop in object space at the axial position of the common focus cancels astigmatism simultaneously for both mirrors, and this is independent of the scale of r_2.

Having shown that for M2, total astigmatism and spherical aberration are both invariant under r_2 scaling (and therefore zero because they were zero when $r_2 = r_1$), the last remaining task is to show that coma is likewise invariant.

Recalling that in the case of a single paraboloid, coma was nonzero and invariant with respect to the stop position, we can set the stop to lie at the surface of M2. In this case, x_{A2} is zero, so the only contributor to consider for M2 coma is that arising from the base sphere. We now only have to look at the definition of coma from equation (2.32):

$$\begin{aligned} \text{coma: } -4\frac{u}{y_c}xW &= -4\frac{u}{y_c}x \times \frac{-1}{4}nc^3y_c^4 \\ &= u.\ x.\ n.\ c^3.\ y_c^3. \end{aligned} \tag{3.7}$$

We have already shown how, as a consequence of Lagrange invariance, as r_2 scales down, W_{S2} scales down with it, while u_p, the chief ray angle, scales up in the same proportion. Because M2 has one infinite conjugate (in image space), it can be considered to be already lying in its own plate space without having to be imaged through M1. In this case, $x_{S2} = r_2$, and so the product $u.\ x$ must be invariant, like the product $c^3.\ y_c^3$ as determined above.

Remember that with our coincident, cancelling paraboloids from the start of this discussion, we see that only the sign of n differs for the two mirrors. Applying equation (3.4) to both shows they must cancel. And now we have seen that as r_2 scales, the cancelling contribution of coma from M2 does not change.

Figure 3.9 shows the plate diagram for the Mersenne system in the case where the entrance pupil is collocated axially with the common focus of the mirrors. The centre of curvature and the surface vertex of M2 are equidistant from the common focus of M1, so when these points are imaged through M1, they are equally dispersed in opposite directions about that point, which itself lies at infinity in plate space. This symmetry means that the astigmatisms of each paraboloid individually cancel themselves. Considering coma (we can compare it to 'torque' about the focus point pivot), we see in this diagram that the M1 contributions are both 'torquing clockwise' (the product of their components is negative), whereas the M2 contributions are both torquing anti-clockwise.

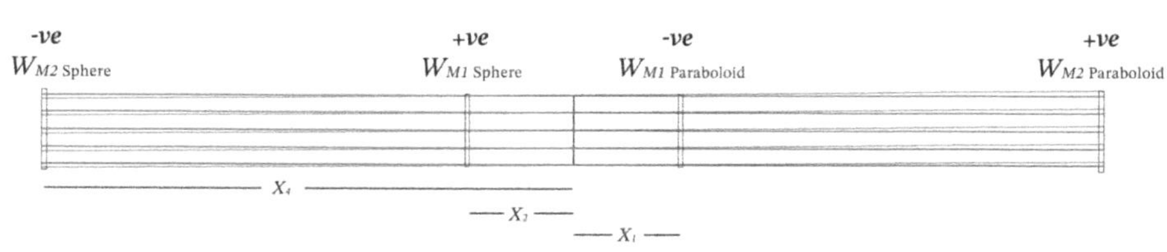

Figure 3.9. The symmetry of the Mersenne telescope plate diagram is clear. Note too that from left to right, the plates alternate in sign, as was proved necessary in section 3.1.

This is an interesting point at which to stop: we have reached the realisation that the Mersenne beam expander is essentially a coma corrector, as the component mirrors are already corrected for spherical aberration and astigmatism. And as the radius of M2 decreases, the plate strengths decrease in direct proportion and the x distances decrease in linear proportion, so the 'torque' remains constant.

3.5 Two-mirror focal anastigmats

Thinking back to the tabulated plate diagram in table 3.2 for the Cassegrain telescope, we can recall that when we modified it to a Ritchey–Chrétien by adjusting the conic constants, the net result was a significant change in coma and a slight increase in astigmatism.

Considering a Mersenne system with the same first-order M1 radius and spacing as that Cassegrain, we see that to make the Cassegrain afocal, the radius of M2 has to increase until it is

$$2\left(\frac{r_1}{2} - t1\right) = -840.521\ \text{mm}. \tag{3.8}$$

At this point, both mirrors become paraboloids. We ask the question, 'Is it possible to make a two-mirror focal anastigmat with the same primary mirror and mirror separation as those we have been using?' It should be possible because we can vary two asphericities and the M2 radius, and we want to control the three aberrations: spherical aberration, coma, and astigmatism.

Table 3.5 modifies the table for the Cassegrain telescope that was derived earlier (table 3.2) to indicate the quantities that need to change to produce a two-mirror focal anastigmat. Three quantities are now variables: the M2 radius and the asphericities of M1 and M2. With these variables, we wish to solve for zero spherical aberration, coma, and astigmatism.

Considering the following expressions:

$$\begin{aligned} W_{S1} + W_{A1} + W_{S2} + W_{A2} &= 0 \quad &\text{(spherical aberration)}\\ x_{S1} W_{S1} + x_{A1} W_{A1} + x_{S2} W_{S2} + x_{A2} W_{A2} &= 0 \quad &\text{(coma)}\\ x_{S1}^2 W_{S1} + x_{A1}^2 W_{A1} + x_{S2}^2 W_{S2} + x_{A2}^2 W_{A2} &= 0, \quad &\text{(astigmatism),} \end{aligned} \tag{3.9}$$

Table 3.5. Quantities involved in the calculation of a two-mirror focal anastigmat.

Quantity	W_i	x_i
M1 S	0.244 141	−11 517.9
M1 A	$-0.244\,141 + 2759.700\dfrac{(\frac{105.0657}{r2} + 0.25)^2}{r2} + \dfrac{3.04638 \times 10^7}{r_2^3}$	−7517.85
M2 S	$2759.700\dfrac{(\frac{105.0657}{r2} + 0.25)^2}{r2}.$	$-\dfrac{(-1579.737\ 17 + \text{r2})}{1 + \frac{-1579.737\ 17 + \text{r2}}{2000}} - 7517.853$
M2 A	$\dfrac{3.046\ 38 \times 10^7}{r_2^3}$	0

we can see that there is a problem. In the expression for spherical aberration, only W_{S1} is known. Therefore, we have three unknowns: the spherical aberration arising from the aspheric term for M1 and both the aspheric and the spherical terms for M2. Also, in the coma and astigmatism expressions there arises one more unknown, x_{S2}, from the unknown radius of M2. This gives a system of three equations in four unknowns.

There is a way around this because the two spherical aberration terms arising from M2 are dependent on two unknown variables, namely the M2 radius r_2 and the M2 conic constant, k_2. Also, x_{S2} is dependent on the unknown variable r_2.

There are several ways to proceed. A relatively simple one is to put the aperture stop at the surface of M2. This eliminates the terms involving W_{A2} from the coma and spherical aberration equations, resulting in the following system:

$$\begin{aligned} W_{S1} + W_{A1} + W_{S2} + W_{A2} &= 0 \quad \text{(spherical aberration)} \\ x_{S1} W_{S1} + x_{A1} W_{A1} + x_{S2} W_{S2} &= 0 \quad \text{(coma)} \\ x_{S1}^2 W_{S1} + x_{A1}^2 W_{A1} + x_{S2}^2 W_{S2} &= 0. \quad \text{(astigmatism)} \end{aligned} \tag{3.10}$$

The centre of curvature of M2 now lies at a distance of $t_1 + r_2$ to the left of M1. Calculating the position of its image through M1, we get

$$x_{S2} = -2000 + \frac{4 \times 10^6}{420.262\,83 + r_2}.$$

As we previously calculated, the image of the vertex of M2 in object space lies 7517.85 mm to the right of M1 (table 3.2). If we now move the stop from M1 to the image of M2 in object space, we subtract this distance from the former values of x_{S1} and x_{A1} and we obtain the new values for the plate positions given in table 3.5. x_{A2} is now zero.

We can now also calculate expressions for W_{A1}, W_{S2}, and W_{A2} in terms of k_2 and r_2. We use again the relation $i = \varphi - u$ to calculate i, noting that $\varphi = \frac{y}{r}$ and the y and u values from the Cassegrain calculation are unchanged and can be substituted here:

$$i_2 = \frac{105.06570}{r_2} - (-0.25). \tag{3.11}$$

Substituting this value for i and y into equation (2.28), and also replacing c_2 with $\frac{1}{r_2}$, we get the following expressions for W_{S2} and W_{A2}:

$$\begin{aligned} W_{S2} &= -\tfrac{1}{4} n_2 c_2 i_2^2 y_2^2 \\ &= 0.25 \frac{(\frac{105.0657}{r_2} + 0.25)^2}{r_2} \times 105.065\,70^2 \\ &= 2759.700 \frac{(\frac{105.0657}{r_2} + 0.25)^2}{r_2}. \end{aligned}$$

Table 3.6. Four solutions, A–D, are found for the system of equations under discussion.

Solution	R1 (mm)	R2 (mm)	k1 (mm)	k2 (mm)	t1 (mm)
A	−4000	−840.52	−1	−1	−1579.74
B	−4000	−420.263	−1.533 34	−0.317 27	−1579.74
C	−4000	0	−1	−1	−1579.74
D	−4000	548.662 08	−4.411 2728	−0.800 6692	−1579.74

$$\begin{aligned} W_{A2} &= k \times -\frac{1}{4} n_2 c_2^3 y_2^4 \\ &= \frac{3.046\,38 \times 10^7}{r_2^3}. \end{aligned} \tag{3.12}$$

W_{S1} is unchanged from its original value of 0.244 141.

Table 3.5 summarises these results.

The system in equation (3.7) is populated with these values and solved simultaneously in Mathematica™. In this code, ‘Solve’ is an exact solver, not numerical, so exact solutions can be found, but the author has not taken the trouble to do so here. Four solutions are found, and these are given in table 3.6.

The first solution is the Mersenne afocal beam expander. Its radius is as expected from equation (3.5), making the secondary mirror confocal with M1. Both conics have gone to −1. We made no stipulation that the solutions should be focal, so it should have been expected that this solution would appear.

Solutions B and C are both very interesting and totally impractical. They are interesting because they should not actually be there. Both are corrected for spherical aberration and coma but exhibit strong astigmatism, both in the Mathematica evaluation and in the ray-tracing code cross-check.

Note that summing the radius of Solution B with the mirror separation t_1 provides a clue about this behaviour. M2 is centred on the focus of M1 in this case. This means that when the plate for the base sphere is projected into object space, it goes to infinity. The chapter on finite conjugate systems has interesting things to say about what happens when plates go to infinity, and this strange result, where coma and spherical aberration are corrected but astigmatism is not, is a main point. It turns out that the plate strength also goes to zero as the plate goes to infinity, and in this case, coma, as the product of zero and infinity, becomes zero, but astigmatism, the product of infinity squared and zero, does not. In fact, it becomes something that depends entirely on the system's Petzval sum. This will be discussed in detail later.

Solution C is another degeneracy: when the radius is zero, curvature is infinite and things break down in a different way. The very surprising thing about both of these degenerate systems, which from the plate-diagram point of view are degenerate in totally unrelated ways, is that the value of the astigmatism residual is exactly the same in both cases.

Solution D is the ‘practical’ focal two-mirror anastigmat we were expecting to find. The layout diagram is shown in figure 3.10, and wavefront coefficients

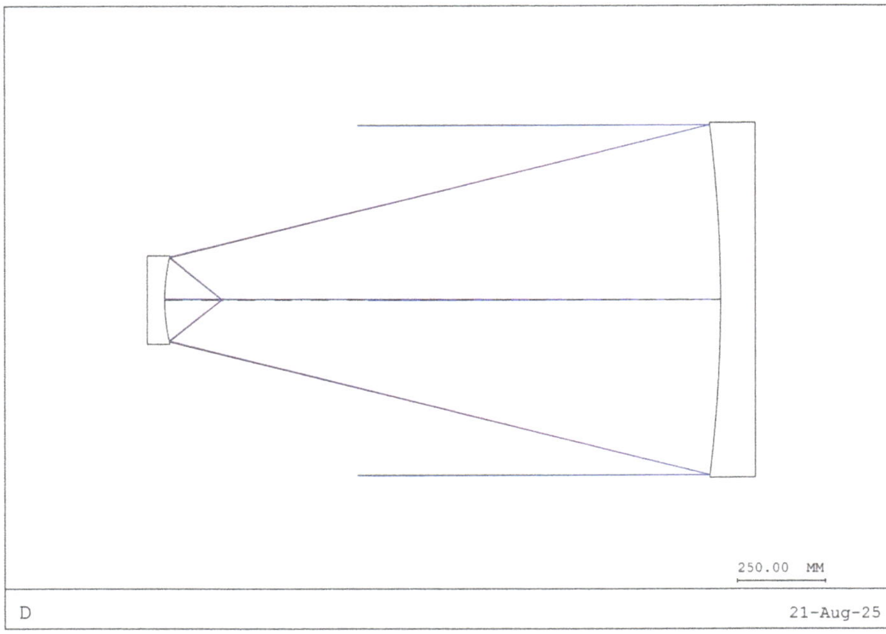

Figure 3.10. Layout diagram for Solution D. Created with Code V.

Table 3.7. Ray-tracing software confirms that the plate diagram solution found is an anastigmat (to rounding errors).

THIRD ORDER IMAGE ABERRATIONS (Wavelengths at 1000 nm)					
Surface	W040	W131	W220	W222	W311
SUM	0.0000	0.0000	-0.7892	0.0000	-0.0194

produced using ray-tracing software are listed in table 3.7. Figures 3.11 and 3.12 are a spot diagram and a Seidel diagram, respectively. Figure 3.13 gives the associated field curvature plots.

As will be described in the next section, this system is one of two types of two-mirror anastigmat first discovered by Schwarzschild in 1905 [4]. The solution we examined here, with a concave primary mirror, was at the time discounted by Schwarzschild because of its strongly curved field, which was not compatible with the flat photographic plates he was working with. He opted instead to reintroduce astigmatism to flatten the field, just as Petzval did with his portrait lens (figure 3.15). André Couder [16], who reported Schwarzschild's concave primary anastigmat in 1926 and proposed a return to the anastigmatic solution, often gets credited with Schwarzschild's discovery. The design lay dormant as a technical curiosity for many years until modern mirror-making technology caught up with optical theory. In modern times, this design has been found to be ideally suited to various types of systems, for example, the Cherenkov Telescope Array [17].

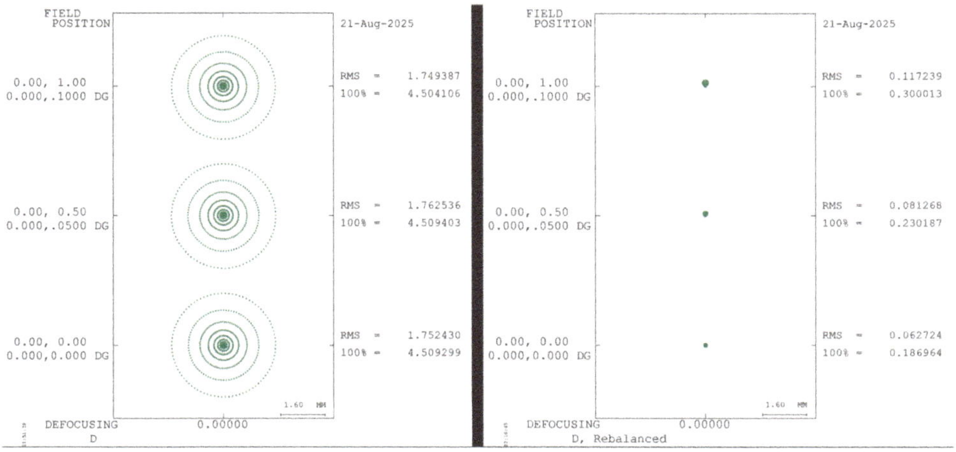

Figure 3.11. Spot diagrams for Solution D (left) and a version where ray tracing was used to reintroduce third-order aberration by just varying the conic constants, to balance against high-order residuals. The layout diagram in figure 3.10 is unchanged in both versions. The same results could be achieved using higher-order aberration coefficients, but that is beyond the scope of this edition of this book. Created with Code V.

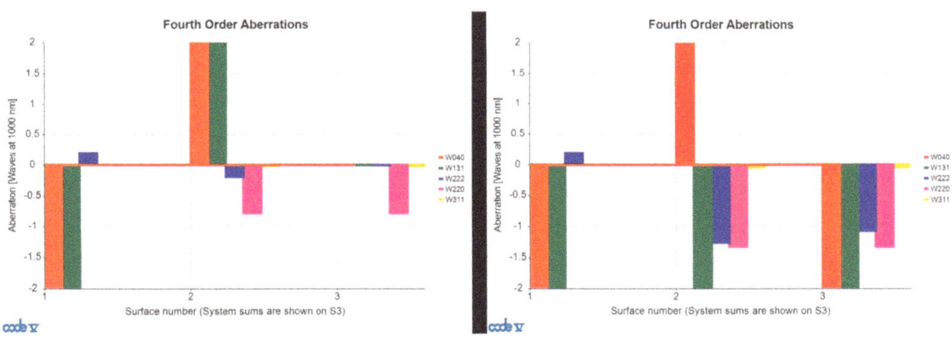

Figure 3.12. Seidel diagrams for Solution D (left) and the rebalanced version (right). We see in Solution D that all Seidel aberrations are zero except for field curvature. Created with Code V.

Now that we have calculated values for a two-mirror focal anastigmat, it is of interest to refer back to an earlier result: the derivation of the four-plate theorem. We see here that the two-mirror anastigmat requires two conicoid mirrors and thus four plates. Referring to figure 3.14 and noting that the system we are interested in has the order of plates indicated in part B of the figure, for a concave mirror, and noting that the stop is on M2, and so x_i distances are measured from the location of W_{A2} in object space, we obtain

$$a = x_{S1} - x_{A1}; b = x_{A1} - x_{S2} \text{ and } c = x_{S2}. \tag{3.13}$$

In fact, as the stop-shift theorem proves later, sign is irrelevant for this exercise, and we will use magnitudes of $a,\ b,$ and c in the following exercise.

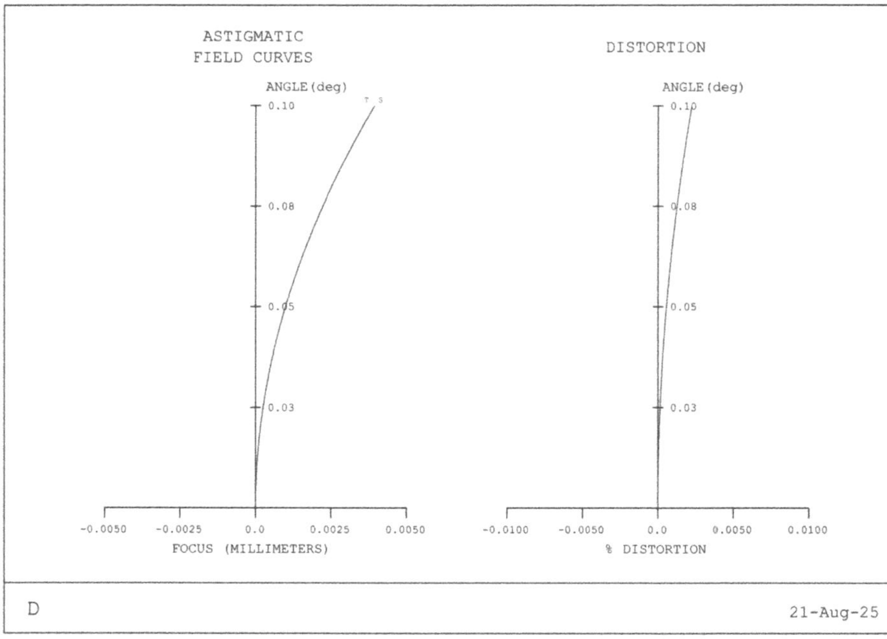

Figure 3.13. Field curvature plots for Solution D. The plot shows zero astigmatism and the focal surface defined by the Petzval curvature. Created with Code V.

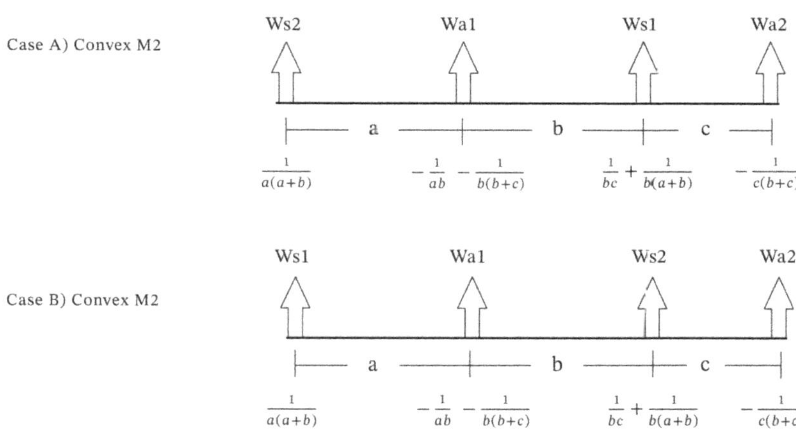

Figure 3.14. Plate diagram for two-mirror anastigmats, showing the order of plates for convex primary mirror systems (top) and concave (bottom). The concave primary Case B matches Solution D analysed in this section. The last row of the table applies a scale factor derived from the first column to the other plate-analysis-derived W_i and shows that the plate-derived W_i are in the same ratio as predicted by the analysis of moments of a mechanical system. Also, $\sum W_i = \sum W_{i\ \text{normalized}} = 0$, as expected.

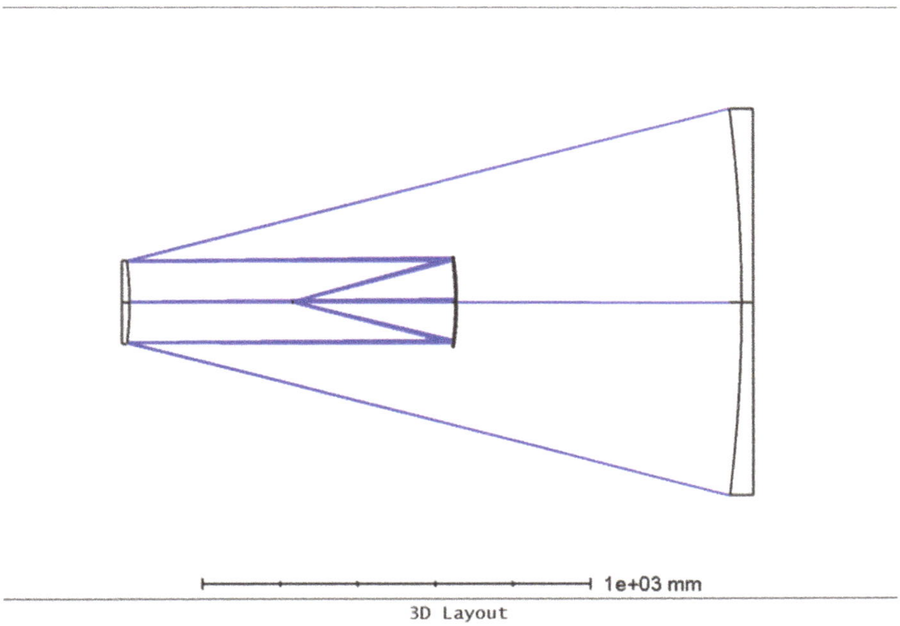

Figure 3.15. A Paul TMA with a paraboloidal primary and spherical secondary and tertiary mirrors. Created with Code V.

Table 3.8. Quantities derived from the plate equations compared to the theoretical ratios derived from moment analysis of a mechanical system.

Quantity\component	M1 S	M1 A	M2 S	M2 A
W_i	0.244 141	−1.076 97	0.980 543	−0.147 713
x_i	−11 517.9	−7517.85	−5389.4	0
a, b, and c substituted into 'Case B' shown in figure 3.14.			$a = 4000$ $b = 2128.4$ $c = 5389.4$	
$W_{i\ \text{normalized}}$ (calculated from a, b, and c)	4.0794×10^{-8}	-1.800×10^{-7}	1.638×10^{-7}	-2.468×10^{-8}
$W_i \times \dfrac{4.0794 \times 10^{-8}}{0.244141}$	4.0794×10^{-8}	-1.800×10^{-7}	1.615×10^{-7}	-2.468×10^{-8}

Table 3.8 lists the plate values for Solution D and relates them to a, b and c.

These results validate the analysis of the mechanical system and present these relations as a powerful tool that can be relied upon to determine mirror radii, aspheric terms, and/or locations, depending on known elements of the system, without the tedium of detailed calculation.

Before proceeding with more technical discoveries, we can take a break for a bit of interesting historical context.

3.6 Historical interlude: legacy lost and found

The two-mirror anastigmat described here is one of two types first discovered by Karl Schwarzschild in 1905 [4]. In the first of his three optics papers of that year [18], Schwarzschild laid down a complete aberration theory to the 5th order of expansion. In this, he used the so-called 'Eikonal' of Bruns [19], which had previously been applied to problems in celestial mechanics. Schwarzschild, being an astronomer, was aware of the Eikonal and decided to apply it to the problem of optical design. But he was also aware of two other great contributors to the theory that he built on: Hamilton [20] and Petzval.

In the 1830s, William Rowan Hamilton (1805–1865) published a book: *Theory of Systems of Rays* [21], in which he developed and used his 'characteristic function' to expand the possible perturbations in pencils of rays. This is the first known example of the expansion of the optical polynomial to produce distinct families of aberrations, but it was not done in a way that could be used for optical design. The expansion did not involve optical system constructional parameters and was not oriented towards producing a corrected optical system. When Heinrich Bruns developed his Eikonal in 1895, he apparently did not realise that he was rediscovering the characteristic function of Hamilton.

In *The Iconic Eikonal and the Optical Path* on the Galileo Unbound blog (November 14, 2023), the blogger David Nolte states: 'Possibly motivated by his studies done with Hausdorff on refraction of light by the atmosphere, Bruns became interested in Malus' theorem for the same reasons and with the same goals as Hamilton, yet was unaware of Hamilton's work in optics' [22].

And so, in 1905, Schwarzschild, already aware of Bruns, Hamilton, AND Petzval, realised that this characteristic function—now called the Eikonal—could be useful in optical design and became the first person *on the available record* to expand the optical polynomial to the fifth order in terms of system construction parameters (useful for optical design) and the first to apply them to the design of telescopes, immediately producing a whole new class of system: the reflecting anastigmat.

I say 'on the available record', but we can reproduce here, from Schwarzschild himself, another specific acknowledgement from an early 20th-century expert of some work of Joseph Petzval's that is no longer available in modern times but clearly was to Schwarzschild. In the opening material of his 'Introduction to Geometrical Optics I', Schwarzschild notes:

> 1. The current report presents a general introduction to the aberration theory of optical systems with the intention on the one hand to give the non-specialist reader a compact overview on the area and on the other hand to produce for my own benefit a source of reference material to be used in future investigations. The representation is based on Hamilton's 'Characteristic Function' which I will name together with Mr. Bruns as the 'Eikonal'.

> ... The number of independent aberrations of the 5th order amounts, without more detail, to 9. Petzval, the calculator of the first 'portrait lens' *gave this number as* 12,[1] from which seems to follow that despite his calculations extending to aberration coefficients of the 9th order, he did not see through the relationship all too deeply.... [18].

In more recent times, de Meijere and Velzel showed that one could obtain either **nine or twelve** independent 5th-order aberration coefficients, depending on the choice of pupil-coordinate definitions [23], thus validating what Schwarzschild saw as Petzval's mistake.

Schwarzschild's genius illuminated the field of optical design, as it did every area to which he turned his attention.

Schwarzschild met a tragic end. He contracted a skin disease while serving on the Eastern Front, while at the same time writing famous papers in physics, including the first exact solution to Einstein's field equations. Schwarzschild died of the disease in 1916 [24].

We now move on to consider anastigmats with more geometrical possibilities.

3.7 Three-mirror anastigmats

The Schwarzschild solutions offered unprecedented (for their day) image quality over wide fields of view, but they suffered from several problems. As noted previously, technological development pertaining to the manufacture of precision aspheres was in its infancy in the early 20th century, and for many years such systems remained interesting theoretical possibilities.

The first Schwarzschild/Couder telescope, a 60 cm version, was made by the University of Indiana in the 1930s [25]. C. R. Burch was enthusiastically hand-polishing Schwarzschild microscope objectives (convex primary solutions) in the 1930s and described his work in detail [26]. The reflecting microscope objectives of Burch represented the state of the art in their day, and coupled with Zernike's phase-contrast method, led to revolutions in biosciences discovery [27].

A geometrical consequence of using only two mirrors is that the solution space is very limited. In the form useful for telescopes (concave primary), the resultant f-numbers of the only solutions are very low (the f-number of Solution D is f/0.8), but the system suffers from a large central obscuration, a difficult-to-access focal surface, and a 'negative telephoto effect', meaning that the tube is longer than the focal length. In fact, Schwarzschild showed that in all cases of two-mirror anastigmats, the separation between the two mirrors is twice the system focal length [4]. This 'tube length = $2f$' rule only breaks down with the Mersenne beam expander, which is afocal.

The first step in broadening the design range of anastigmats made solely from reflectors was taken by Maurice Paul [28]. Paul produced the world's first three-mirror anastigmat, which is described below. Soon after, the same form was

[1] Author's emphasis here.

independently rediscovered by Baker and published in 1940 [29]. Interest in this telescope form, and others with four or more mirrors, exploded from there, and describing them all would fill a book on its own. In particular, from the late 20th century, three-mirror anastigmat (TMA) designs have increasingly been used in space applications, where the combination of a strong telephoto effect, wide field, compactness, and excellent correction has given them a dominant position.

In his 2000 master's thesis, the author made the sort of sweeping grand statement one can often find in M.Sc.s: 'If the Twentieth Century can be thought of as the century of the Ritchey–Chrétien telescope, the Twenty First Century will be the century of the TMA' [30]. Now, 25 years later, we see TMAs dominating space, with the James Webb telescope as a flagship for this class [31]. TMA design is ubiquitous in space missions, given the design's inherent advantages in 'étendue per unit volume' and likewise mass. On the ground, the European Extremely Large Telescope (ELT), which is a TMA, will be the largest optical–infrared telescope humankind has ever produced when it sees its first light in the early 2030s [32]. Also, at the time of writing, the newly commissioned Vera C. Rubin telescope is providing unprecedented étendue [33] (defined as the product of the clear aperture and field areas, as in table 3.9).

Before moving on to an investigation of TMA plate diagrams, in tribute to the author's mentor, Norman Rumsey [34], one innovation in TMA design will be mentioned. In 1969, Rumsey considered that, with the available degrees of freedom in a TMA (eight degrees of freedom: three radii, three conicoids, and two air spaces) and the desire to control four of the five Seidel aberrations (discounting distortion, as is common), focal length, and focal plane position, there would be two degrees of freedom left over. What to do with these? Why not, Rumsey posited, arrange the design so that the outer edge of the tertiary mirror and the inner edge of the hole in the primary mirror coincide both in space AND in slope? In this way, two different conicoid surfaces could be made on a continuous optical surface. Rumsey pointed out the natural alignment advantages of having the tertiary mirror physically coupled to the primary mirror in this way [35].

The Rubin telescope makes good use of this alignment advantage, with M1 and M3 made on the same substrate (but not slope matched) [36]. Often described as a 'modified Paul' telescope, it is more accurately described as a 'modified Rumsey' telescope, as the key innovation of manufacturing the tertiary and the primary on a common substrate, in addition to the associated variation in conic forms and first-order geometry from Paul's design, were first proposed by Rumsey. Also, in the early 2000s, Gerard Lemaître, pioneer of the Reflecting Schmidt Telescope [37, 38], as embodied by the Large Sky Area Multi-Object Fiber Spectroscopic Telescope (LAMOST) [39], has developed the Rumsey design for space applications [40].

For our plate-diagram investigation, we will start at the beginning, with the Paul telescope.

The plate-diagram investigation of this telescope is very simple. At the beginning, we must recall some of the results from the last chapter: in particular, the concepts of the Schmidt telescope and the paraboloid mirror, along with equation (2.35)

Table 3.9. Comparison of étendue for various existing telescopes. The new Vera C. Rubin TMA clearly dominates. Before that, Lemaître's reflecting Schmidt anastigmat dominated for more than a decade.

Rank	Telescope	Aperture (m)	Field of view	Telescope type	Étendue ($m^2 \cdot deg^2$)
1	Vera C. Rubin observatory (Large Synoptic Survey Telescope, LSST)	8.4	3.5° diameter (9.6 sq deg)	Three-mirror anastigmat	319.5
2	Large Sky Area Multi-Object Fiber Spectroscopic Telescope (LAMOST) (Xinlong observatory)	4.0	5.0° diameter (~19.6 sq deg)	Two-mirror Schmidt	185.0
3	Dark Energy Spectroscopic Instrument (DESI) (Mayall telescope)	4.0	8.0 sq deg	Corrected prime focus	100.5
4	4-metre Multi-Object Spectrograph Telescope (4MOST) built for the Visible and Infrared Survey Telescope for Astronomy (VISTA)	4.1	4.2 sq deg	Corrected Cassegrain	55.5
5	VISTA	4.1	1.65° diameter (2.1 sq deg)	Quasi–Ritchey–Chrétien	28.2
6	Panoramic Survey Telescope and Rapid Response System (Pan-STARRS)	1.8	3.0° diameter (7.1 sq deg)	Corrected Ritchey–Chrétien	18.0
7	Fiber Multi-Object Spectrograph (FMOS) (Subaru)	8.2	30′ diameter (0.20 sq deg)	Corrected prime focus	10.4
8	SkyMapper	1.35	5.7 sq deg	Modified Cassegrain	8.2
9	VST (Very Large Telescope survey telescope)	2.6	1° × 1° (1.0 sq deg)	Modified Ritchey–Chrétien	5.3

($W_A = -k\frac{1}{4}nc^3y^4$), which told us that the spherical aberration of a spherical mirror in collimated light is exactly cancelled by the paraboloidal departure from a sphere, represented by the conic constant, k, of value -1.

With these things in mind, no calculation will be necessary. We can consider the layout of the Paul system, given in figure 3.15.

The optical prescription for this system is so simple that it requires no table. The primary mirror, secondary mirror, and mirror separation of the Paul can have identical values to the two mirrors of our Mersenne system, which we saw most recently as Solution A in table 3.6 in the preceding section. The sole difference is that in the Paul, the secondary mirror is spherical, whereas in the Mersenne it is a paraboloid.

To the first order, the first two mirrors of the Paul are identical to those of the Mersenne, so the first two mirrors make an afocal system; however, by removing the parabolisation of M2, we unbalance the plate diagram, and aberrations of the convex sphere in collimated light arise, causing coma in the near field and astigmatism dominating the outer field.

We can recall that the third-order aberrations in a multi-element system are simply additive. We have used this property all the way through this book so far. This means that we can do some 'anastigmat algebra', by which we mean 'anastigmat + anastigmat = anastigmat'.

If we cojoin two anastigmatic systems, the resultant system is anastigmatic. So, consider now a Schmidt telescope that takes the output from a Mersenne telescope. The Mersenne is anastigmatic and produces collimated light. The Schmidt is anastigmatic, is fed with collimated light, and produces an image on a surface having half the radius of curvature of its powered mirror.

The sum of these two systems is also anastigmatic.

If we arrange things so that the concave radius of the Schmidt mirror has the same magnitude as the convex radius of the Mersenne M2 mirror, then by equations (2.29) and (2.35), we can see that the parabolising plate for the Mersenne secondary is of exactly the same magnitude and opposite sign to the Schmidt plate in this case, so they exactly cancel. Further, if we locate the Schmidt telescope axially, so that the centre of curvature of its spherical mirror is located at the vertex of the paraboloid mirror, then the Schmidt plate and paraboloidal figure plate are co-located and cancel exactly.

We are left with the Paul TMA system, a version of which Roderick Willstrop referred to as the 'Mersenne–Schmidt' [41].

Note the utility of plate-diagram thinking here. Without any extra maths beyond foundational concepts, we can arrive at the third-order solution of a three-mirror system with one asphere!

In figure 3.16, we see the remarkable truth about the 'Mersenne–Schmidt'. The plate diagrams for the two systems, the Paul TMA and the Mersenne, are identical, with the same plate strengths and the same plate positions. The only difference in the diagram is the label for the plate on the far right. Where previously that contribution came from the parabolisation on M2 of the Mersenne, it now arises from the tertiary

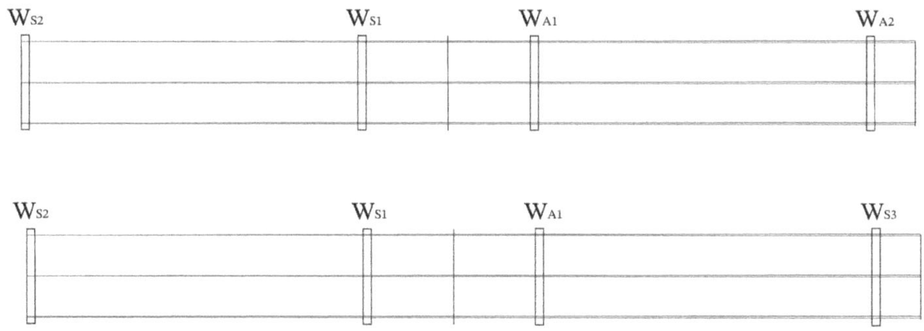

Figure 3.16. Plate diagrams (with distances scaled) for the Mersenne system (top) and the Paul TMA (bottom). The plates for the two systems have identical locations and strengths. The unscaled distances for both plate systems, together with correctly calculated weights, give a special degenerate case of the moment-balancing diagram, where in this case $a = c$.

spherical mirror of the Paul. With its centre of curvature on the M2 vertex, this plate images to exactly the same location in object space as the plate from M2 aspherisation did.

This highlights visually what has already been stated and also implied mathematically. The plate diagram is not sensitive to whether a contributing 'force' arises from a spherical mirror or an aspheric figure. Any plate diagram can, in fact, represent two or more real physical systems, provided the geometry of each physical system is arranged to put the plates of the same strength in the same places.

One difference between these two systems, which is not captured in the plate diagram, is the Petzval curvature. Remembering that for mirrors in air, from equation (2.33) we have $\sum_{i=1}^{m}\frac{2}{r_i}$, and considering that $r_2 < r_1$ and usually less than half of M1's radius, we see that the Mersenne has a field curvature dominated by its small secondary mirror.

On the other hand, with the Paul TMA, the additional concave mirror exactly cancels the curvature of the convex mirror, and so the system field curvature is reduced from that of the Mersenne to now equal $\frac{2}{r_1}$.

Baker, mentioned earlier as having independently rediscovered Paul's invention by 1940, described a variant of the Paul TMA in 1969 in which he completely flattened the field. He realised that the residual Petzval was now weakly concave, originating from the uncancelled curvature of M1. By reducing the total concave curvature, achieved by increasing the length of the M3 radius, he introduced enough positive curvature to cancel the M1 curvature, thereby producing a flat field. He corrected the aberration that resulted from this adjustment by making M2 slightly ellipsoidal. In his paper, Baker worked out analytically exactly how much adjustment would be needed.

Armed with the plate diagram, we can go immediately to this solution. If we consider our Paul TMA again to be two anastigmatic systems, the Mersenne and the Schmidt, we can see that the Schmidt mirror can have any radius, with the required plate strength changing as r_3^3, and the summed system is still an anastigmat. If we now arrange things so that the Schmidt plate lies exactly on M2 and we choose a radius for the Schmidt mirror such that the Petzval sum of the three mirrors is exactly zero, i.e.

$$r_3 = \frac{r_1 r_2}{r_1 - r_2}, \tag{3.14}$$

then the resultant reduction in plate strength from the Schmidt would mean that the paraboloidal plate on the Mersenne M2 is no longer fully cancelled. Reintroducing figuring in the direction of a paraboloid ($-1 < k_2 < 0$) results in the ellipsoidal figuring Baker introduced. While r_3 has changed, the location of its centre of curvature has not, so the plate representing M3 stays exactly where it was in figure 3.16. The plate strength is also identical, but now it is a summed plate, from the two coincident plates representing the weakened M3 plate and the reintroduced asphericity on M2. The sum gives exactly the same plate strength as was the case for the Mersenne W_{A2} plate, or for the Paul W_{S3} plate. The required value for k_2 can easily be calculated once equation (3.11) is solved, by substituting the resultant r_3 into equation (2.29), then working out the required value for k_2 with equation (2.35) such that the summed plates $W_{S3\text{New}} + W_{A2}$ equal the original W_{S3}.

This system with coincident plates, which we will look at in more detail in the following chapter when we deal with concentric systems, illustrates an interesting analogy. While in real optical systems the optical elements cannot occupy the same space, in plate space any number of plates can overlap. In another way of looking at it, real optical elements behave like fermions, and plates can behave like bosons.

It is a remarkable result that the three optical systems shown in figure 3.17, whose invention spanned a period of 333 years, have identical plate diagrams. There seems to be unexplored space for the development of a visual design software tool here. If one started from the plate diagram and then looked at all the combinatorial possibilities of real optical systems that could give rise to the same plate diagram, unexpected solutions might be found. In the case just discussed, the Paul and Paul–Baker systems would have immediately emerged from a plate diagram for the Mersenne system.

There is one other flat-field anastigmat of similar form to the Paul–Baker, which has a slightly different plate diagram from the three designs discussed so far. The 'Paul–Rumsey' telescope was discovered by the author during a survey that will be discussed further in a following chapter of this book [42]. This telescope maintains the two spherical mirrors of the Paul telescope, but the M3 has the same longer radius that is required to flatten the field as that of the Paul–Baker. The missing spherical aberration correction that occurs when the Paul M3 radius is lengthened to the Paul–Baker radius is now applied to the paraboloid primary mirror, leaving M2 spherical. This necessitates making the paraboloid an ellipsoid, with a conic constant close to -1. These changes require a rebalancing of the plate diagram, but this is achievable by moving M3 back a little, so its centre of curvature now lies above the M2 vertex.

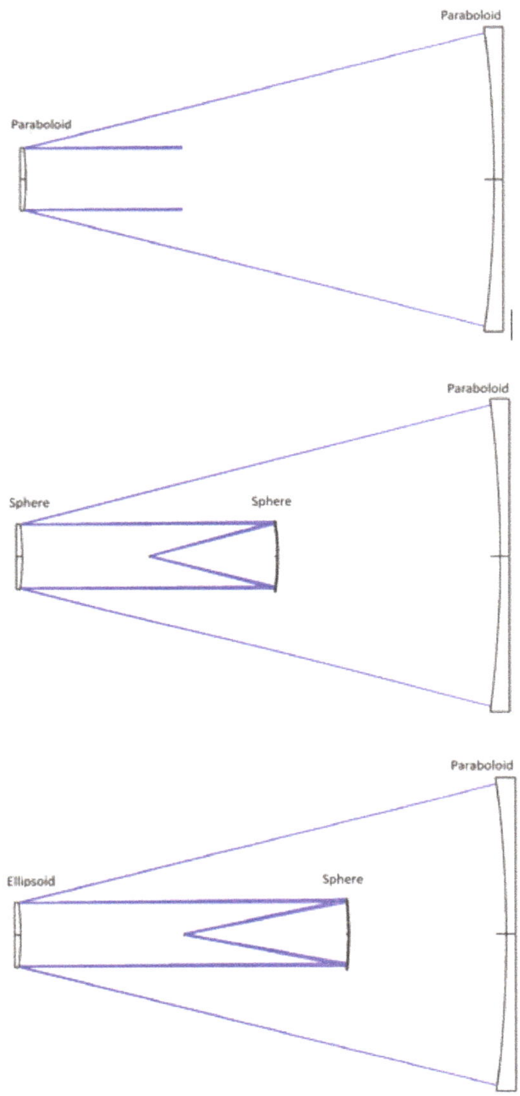

Figure 3.17. Three anastigmats with identical plate diagrams. Mersenne (top), Paul (middle) and Paul–Baker (bottom). Created with Code V.

When the author told his mentor about this design, Rumsey pulled a card from a dusty filing cabinet that had the same exact design written up. He had always intended to write a paper about it but got distracted. Hence the name.

An advantage of the Paul–Baker design is that it can be retrofitted to any existing two-mirror telescope that has a paraboloidal primary mirror, correcting both the coma and the astigmatism of the classical form. The Paul–Rumsey flat-field TMA would require a custom-made primary mirror.

If the radius of M2 (and therefore also of M3) in these design forms is increased, the central obstruction increases, and both M3 and the focal surface move back towards, or even through, a hole in the primary mirror. Roderick Willstrop proposed a form of the Paul system in which M2 and M3 were equally disposed about M1, with the focal surface in the middle of the M1 hole. This system has ~50% linear central obscuration. The Paul–Baker achieves the same form with less central obscuration and the Paul–Rumsey with still less.

One might ask the question: 'What happens if we leave M1 as a paraboloid, M2 spherical, flatten the field with M3, and correct spherical aberration by figuring the concave M3?' It is generally easier to manufacture a concave asphere than a convex asphere of the same radius and aperture.

The answer is that it becomes impossible to fully balance the plate diagram to produce an anastigmat in this case. A flat-field aplanatic solution, correcting spherical aberration and coma, is possible, and plate balancing leads to an interesting solution. In figure 3.18, we see this variant, where M3 is now an oblate spheroid of conic constant +1, and the image forms on the M2 vertex, which is a significantly more convenient location than with the other forms (small modifications can push the image slightly behind a hole in M2, making it accessible without necessarily adding to obscuration with a camera and mount). The residual astigmatism can be corrected and a fully balanced plate diagram achieved with the addition of a very small Gascoigne plate near the focus, giving a flat-field anastigmat. Small modifications to this design can also be made to work with the primary mirror of any Ritchey–Chrétien telescope. Such a design provides an interesting option as a wide-field corrector for a large number of existing telescopes. Corrected fields

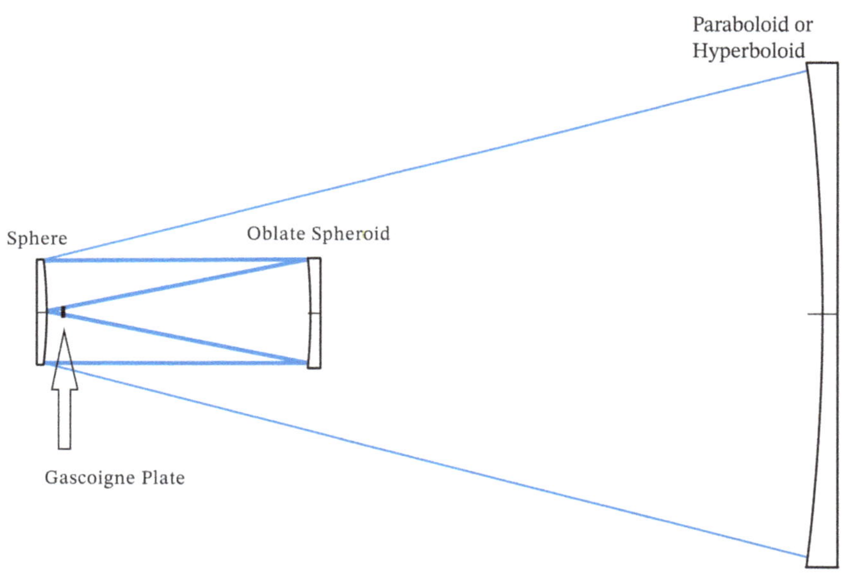

Figure 3.18. An aplanatic flat-field plate-diagram solution can be made anastigmatic by adding a Gascoigne plate. The plate can be moved so close to the focal surface that it could, in fact, be an aspherized Dewar window on a detector.

of over one degree in diameter are achievable without adding significantly to the central obscuration shown here.

We conclude this section by solving a plate diagram for a different type of TMA. Here, instead of using historical examples and plate-diagram-based intuition, we will produce a system layout that is potentially interesting in modern times and calculate a solution.

In this exercise, we will set up and solve the plate equations for a system of three spherical mirrors that have already been set up as a first-order solution. The plate equations will be used to solve for three conics to make the system an anastigmat.

We start with a one-metre diameter flat-field solution with a two-degree field working at f/2.4. A first-order starting solution can be produced in any way most convenient. Ray-tracing software is, of course, commonly used. The enthusiastic student can try a first-order solution. After entering the primary mirror radius and setting the initial height to 500 mm and the initial ray angle to zero, a first-order ray-tracing matrix for a marginal axial ray can also be used to generate equations that can be solved for some combination of specified mirror locations, marginal ray heights, and focal plane position. Solving these equations while simultaneously enforcing the Petzval condition will yield any desired first-order layout, such as the one shown below in figure 3.19. Parameters for this starting solution are given in table 3.10.

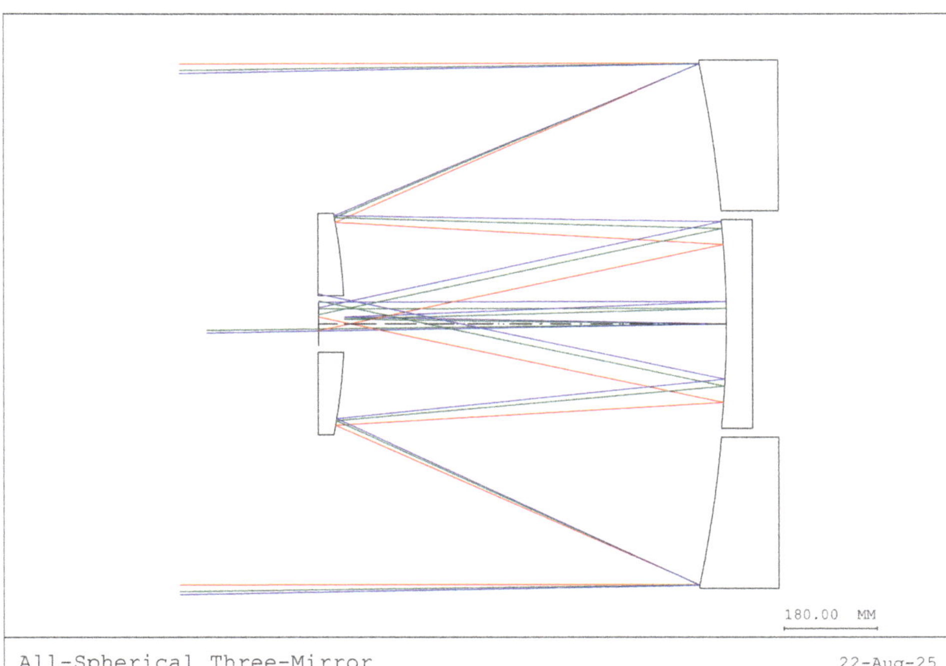

Figure 3.19. First-order layout for a TMA system to solve. All mirrors are spherical at this stage. The system can be seen to have large spherical aberration, as the marginal rays are coming to focus significantly before the paraxial focal surface. Created with Code V.

Table 3.10. Set parameters for the initial system of spherical mirrors. Petzval curvature is zero.

Parameter	Value (lengths in mm)
M1 diameter	1000
Initial stop position	M1
F-number	2.4
M1 radius (r_1) (concave)	−2450.000
M2 radius (r_2) (convex)	−1108.964
M3 radius (r_3) (concave)	−2026.016
M1–M2 distance (t_1)	−735.088
M2–M3 distance (t_2)	735.088
M3 focal distance (t_3)	−784.201

Table 3.11. Calculated plate-diagram quantities. Note that the entrance pupil is on M1, so $x_{A1} = 0$.

Mirror\plate quantity	x_{Si}	W_{Si}	x_{Ai}	W_{Ai}
M1	−2450	1.06248	0	W_{A1}
M2	−3649.069	−0.46797	1838.050	W_{A2}
M3	−4338.017	0.05591	7407.068	W_{A3}

The chosen starting solution combines the attractive features pointed out by Rumsey in 1969 [43]: an accessible image behind M2 and M3 collocated with M1, but no condition is imposed such that M1 and M3 must form a continuous surface.

We shall proceed to set up the plate equations required to solve for the conic constants. The first step is to establish the marginal axial ray heights and incidence angles at M2 and M3. Together with the given values for M1, these are used to evaluate W_{S1}, W_{S2}, and W_{S3}.

Second, the centres of curvature of M2 and M3 must be imaged into object space, producing values for x_{S2} and x_{S3}.

Third, we need to image the vertices of M2 and M3 into object space, producing values for x_{A2} and x_{A3}.

Rather than write all these steps out here, for the benefit of people new to ray-tracing equations, these equations and the results are produced in full in Mathematica script in Appendix 1. They are easy to set up in Mathematica code, with the advice to be careful with sign conventions and when applying refractive indices.

The results of the three calculation steps above are given in table 3.11.

With the values in table 3.12 evaluated from the known quantities of the system, we can set up plate equations and solve for the unknown W_{Ai}, noting that in the

Table 3.12. Fully populated plate-diagram table.

Mirror\plate quantity	x_{Si}	W_{Si}	x_{Ai}	W_{Ai}
M1	−2450	1.062 48	0	−1.51119
M2	−3649.069	−0.467 97	1838.050	0.940 526
M3	−4338.017	0.055 91	7407.068	−0.079 756

coma and astigmatism, W_{A1} is dropped; the reason for this is that because the stop is on M1, therefore x_{A1} is zero:

$$W_{A1} + W_{A2} + W_{A3} = -\sum W_{Si} \quad = -0.650\,424$$

$$x_{A2} W_{A2} + x_{A3} W_{A3} = -\sum x_{Si} W_{Si} = -1137.97$$

$$x_{A2}^2 W_{A2} + x_{A3}^2 W_{A3} = -\sum x_{Si}^2 W_{Si} = 1198\,337. \tag{3.15}$$

Solving for W_{Ai} (using full numerical accuracy and the Mathematica™ code in Appendix 1) gives

$$W_{A1} = -1.51\,119,$$

$$W_{A2} = 0.940\,526,$$

$$W_{A3} = -0.0797\,569.$$

In the final step, we evaluate the three-mirror's conic constants, k_i, using

$$W_A = -k\frac{1}{4}nc^3y^4,$$

which was given previously (equation (2.35)) and the quantities n_i, c_i, and y_i that were already calculated in Steps 1–3 above.

The resultant conic constants are:

$$k_1 = -1.422\,322\,476,$$

$$k_2 = -3.209\,039\,294,$$

$$k_3 = -3.577\,356\,23.$$

The resultant system is shown and described in figures 3.20–3.23. The slightly lower precision required for k_3 can be understood by considering the relative amount of spherical aberration contributed by each mirror to the system sum (figure 3.22).

It is interesting to compare the Seidel diagram to the information we get from the plate diagram, now that we have solved for all 12 values in this case. Consider table 3.12.

The W_S and W_A terms on M1 and M2 are doing most of the work for spherical aberration, leaving uncancelled spherical aberration on each mirror, with signs that cancel out between the mirrors. Conversely, M3 is not doing much with spherical aberration, but its plates at large distances make a difference to field aberration,

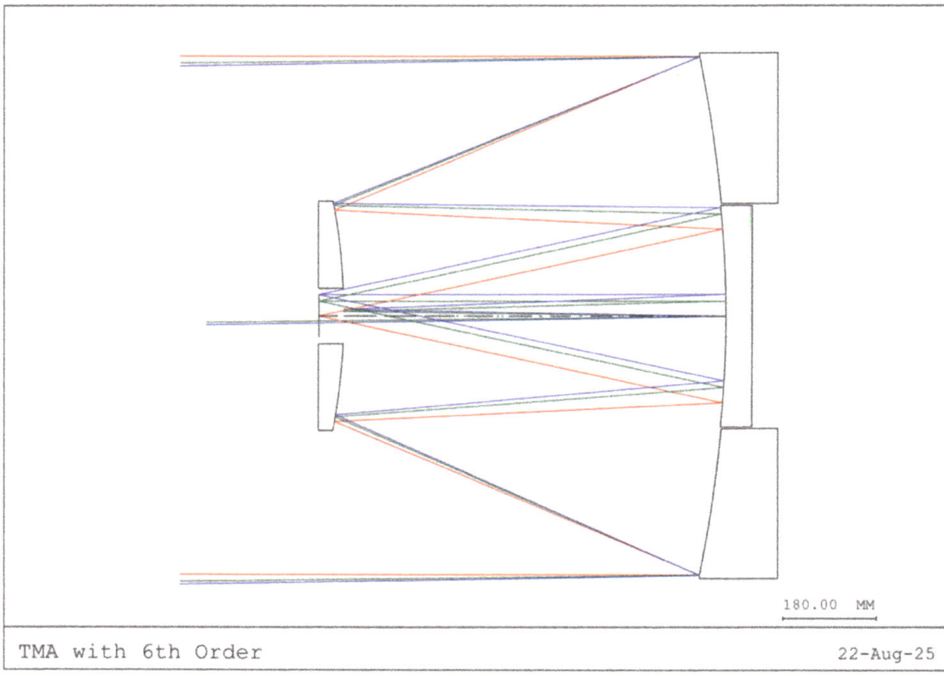

Figure 3.20. The layout diagram now shows marginal rays focusing at the paraxial focal surface. Changing the aspheric terms made no other visible change. Created with Code V.

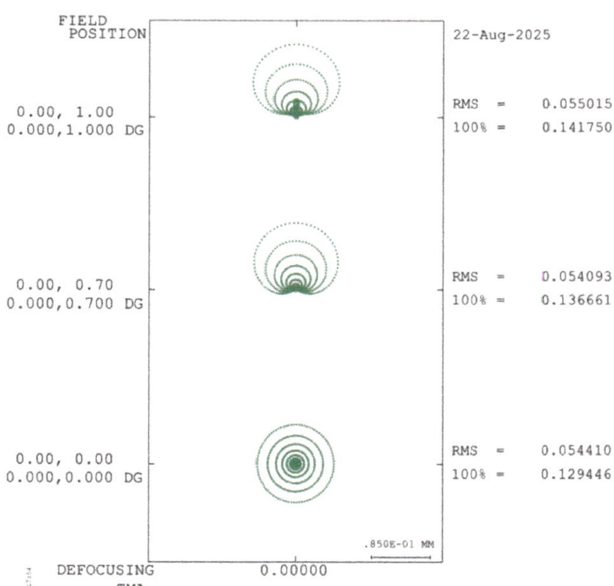

Figure 3.21. A spot diagram shows residual aberration, which, if our calculations are correct, will all be from high-order aberration. Created with Code V.

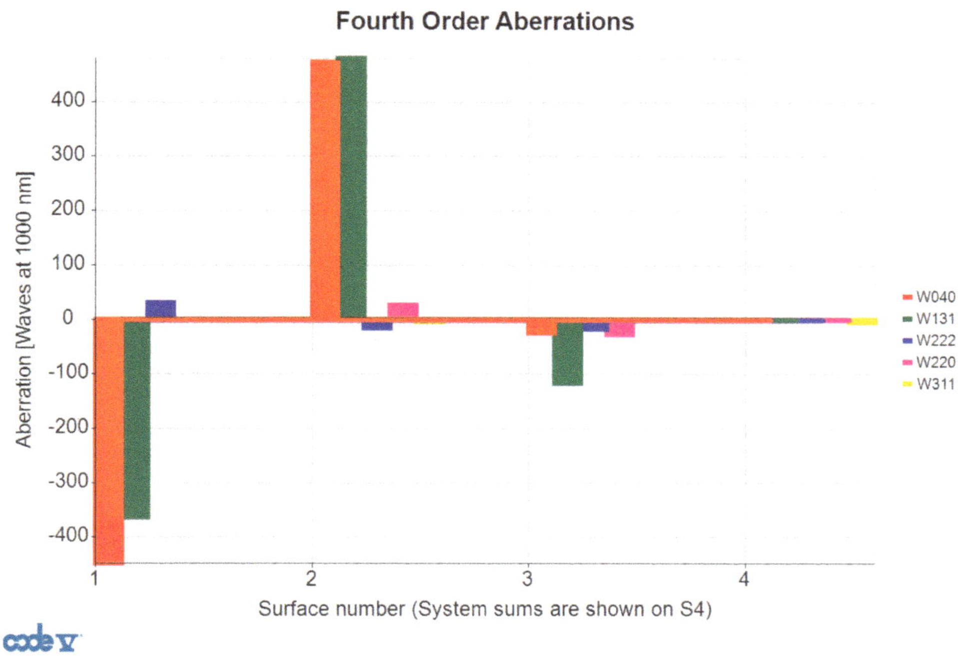

Figure 3.22. The Seidel diagram is interesting. We see that aberration balancing is mainly achieved by M1 and M2, with M3 supplying a very small percentage of the total aberration. The system sum is zero (including for distortion, which was not targeted). This confirms that the aberration in the spot diagrams is all residual high-order aberration. Created with Code V.

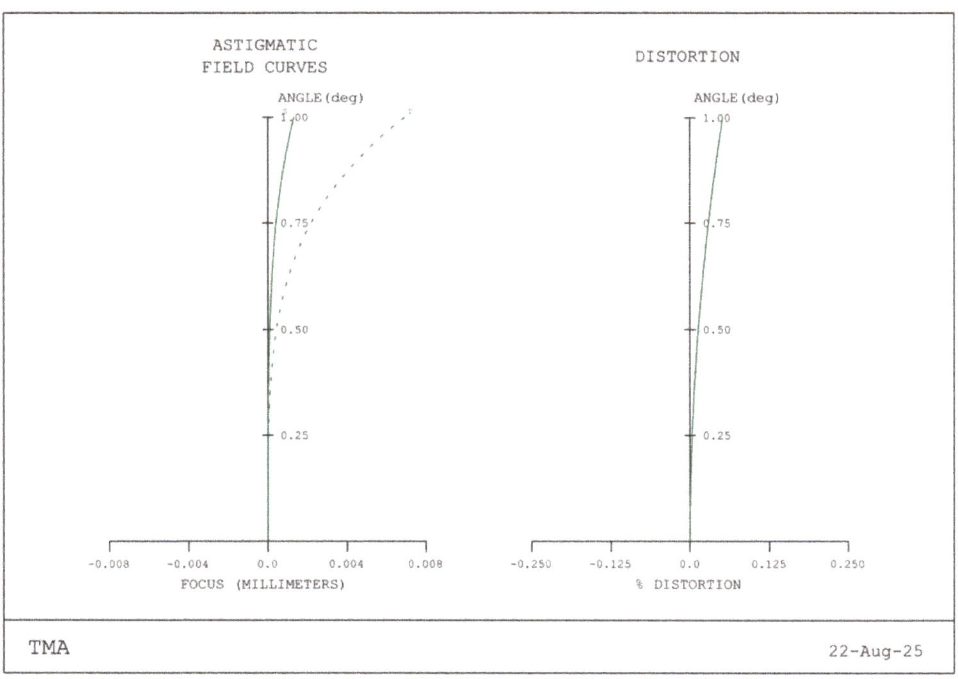

Figure 3.23. The field-curvature/astigmatism plot shows that high-order aberration begins at about 40% of the field height. Created with Code V.

especially coma, where the aspheric and spherical terms produce the same sign of coma when multiplied by their x values. We see this pattern holds with each mirror, with M1 providing net negative coma, M2 net positive, and M3 net negative, and that is what we see on the Seidel diagram. The plate diagram allows us to directly see the separate contributions from the aspheric and spherical terms, which has proved to be useful in the author's experience, both in finding improvements in designs and in identifying 'hard-working' components visually.

As a final point, we note that this 1-metre f/2.4 system working at 1 micron can be brought to the diffraction limit over its two-degree field by allowing a few microns of 6th-degree deformation to be added to M1 and M2. With modern manufacturing technologies and holograms for testing, such mirrors would present no noticeable extra difficulty in fabrication over the pure conicoid forms. Figure 3.24 illustrates the performance improvement over that of figure 3.21.

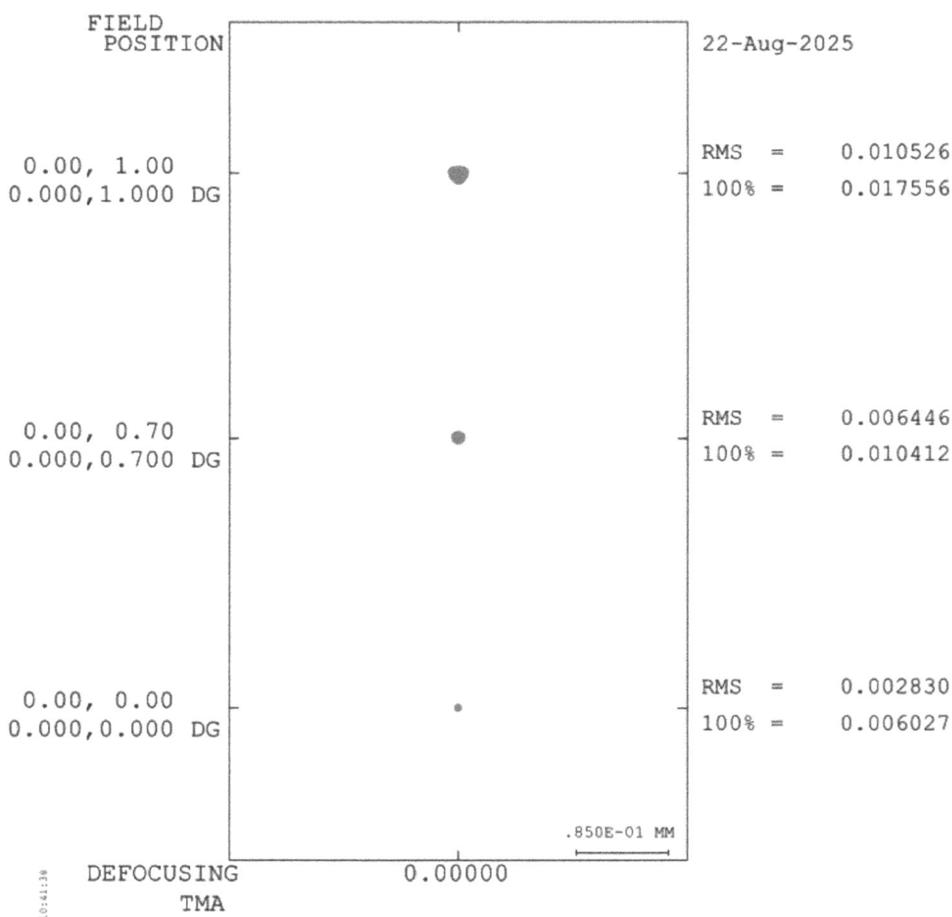

Figure 3.24. The system becomes diffraction limited when we allow a few microns of 6th-degree polynomial deformation to be added to M1 and M2, without changing any parameters at all in the third-order solution. Created with Code V.

This section was not intended to give a thorough survey of TMAs, existing or possible. Rather, the goal has been to illustrate the plate diagram in a few interesting cases. In the author's opinion, an excellent survey of TMA designs up until 2007 can be found in Ray Wilson's *Reflecting Telescope Optics I* [2]. Also, Gerard Lemaître's book contends with Ray's for historical depth and technical breadth [44]. Of course, papers abound about this system and others.

The calculations in the last example were relatively straightforward. The first-order calculations are elementary, widely known, and essentially simple 2D linear matrix operations. Once these first-order quantities are obtained, the full plate solution emerges in a way that seems, at least to the author, more elegant and suited for intuitive understanding and automated solution finding than the denser 'brute-force' algebraic approaches one finds in the literature, and most certainly better for exploring high-dimensional solution spaces than brute-force ray tracing. In a subsequent chapter, this last point will be illustrated with some concrete examples.

The insights that come along the way, which show how the spherical and aspheric terms balance aberration, are valuable in that they help a designer build up an intuitive grasp of the topic, which will not come as easily with ray-tracing-based design alone. This may seem like a big claim, but the next chapter will provide some 'big proof'.

Before leaving this chapter, we will conclude with a very brief discussion regarding systems of more than three mirrors.

3.8 Four and more mirrors

So far, we have seen examples of systems that inherently balance aberration using internal symmetries, such as the Mersenne, Paul, and Paul–Baker systems. We have also seen other systems where less symmetrical compensation is effectively applied, such as the Ritchey–Chrétien and the modified Rumsey telescope from the last example.

Symmetry is a powerful concept in optical design, as in many branches of physics and engineering. A particularly powerful type of symmetry is monocentricity, where optical components share a common centre. In particular, when we consider monocentric spherical systems, we see that we have a case where two or more surfaces give rise to two or more plates that have that 'boson-like' behaviour, i.e. occupying the same space and summing their powers. In some special cases, this produces an anastigmatic system. We have seen two examples of this so far. The Schmidt system, where the plate from the spherical mirror is perfectly cancelled by the coincident Schmidt plate, leaves a plate diagram with zero weight and nothing to balance. Also, we saw how Baker's system used coincident plates arising from the M2 vertex to allow field flattening.

An early four-mirror telescope design made use of monocentricity, together with the 'anastigmat algebra' mentioned above. Burch described this system in a talk he gave in 1973 [45], but it was clear at that point that he was talking about old work. He produced a four-mirror anastigmat by combining two known anastigmats. At the front end, he used the now very familiar Mersenne confocal paraboloidal pair—

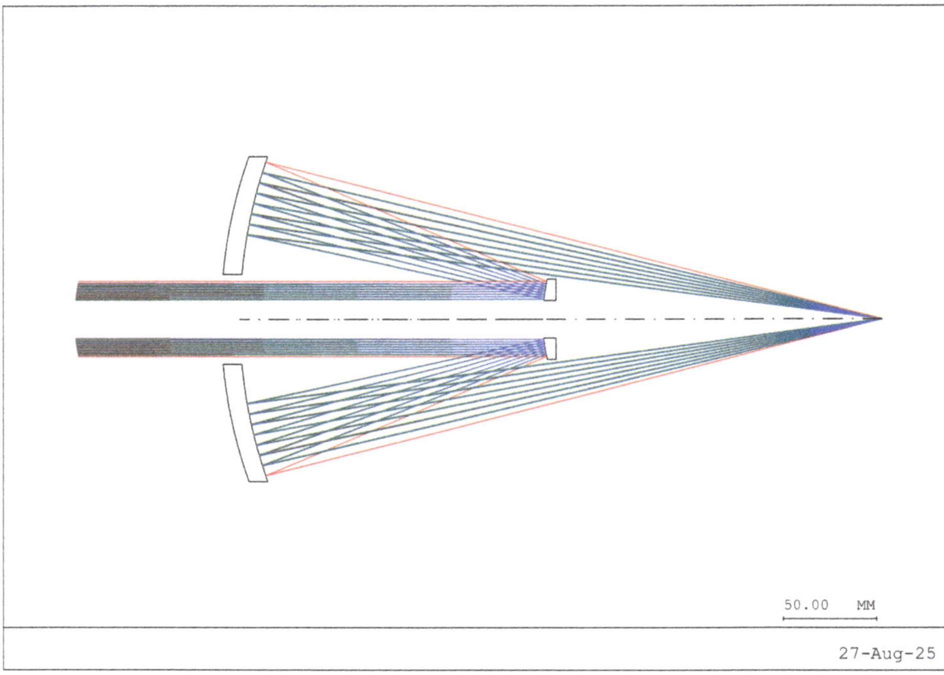

Figure 3.25. The concentric spherical-mirror anastigmat of E.M. Brumberg (1943). Created with Code V.

an 'anastigmatic feeder', as he put it. At the rear of this, behind the primary mirror of the Mersenne, he placed another two-mirror anastigmatic telescope, and this telescope comprised two spherical mirrors.

In fact, this optical system is more commonly referred to as a reflecting microscope objective, which is the purpose for which it was designed. The two mirrors are concentric; the primary mirror, receiving collimated light, is convex and diverges light to the larger concave mirror. An example is shown in figure 3.25. A fun paper by Willard, 'The Golden Ratio in Optics' [46], describes this kind of system and references Kingslake [15] as pointing out that in the anastigmat with two concentric spherical mirrors, the ratio of the primary mirror radius to the secondary mirror radius is given by:

$$\frac{r_2}{r_1} = \phi^2, \quad \text{where } \phi = \frac{1}{1 + \frac{1}{1 + \frac{1}{1 + \dots}}}, \tag{3.16}$$

which is the golden ratio. In fact, the golden ratio is present throughout this system, since we also have:

$$\frac{t_1}{r_1} = \frac{r_2}{t_1} = \phi,$$

and also, the ratio of the height of the marginal ray converging to focus to the height of that ray at the primary mirror is ϕ. Knowing that

$$\phi = \sqrt{\frac{\sqrt{5} + 1}{\sqrt{5} - 1}} \tag{3.17}$$

makes this yet another of our relatively simple 'building block' anastigmats to set up.

In the example given in figure 3.25, M1 has a radius of 100 mm, which is $100(\phi^2 - \phi)$; therefore, the secondary mirror has a radius of $100\phi^2 = 261.8$ mm, and the mirror separation is 161.8 mm.

As a side note, the inventor of this system was very difficult to track down, as almost everyone in the literature refers to this concentric spherical-mirror anastigmat as a 'Schwarzschild anastigmat'. While this solution was implicit in Schwarzschild's equations, he did not notice it; or, if he did, he made no mention of it. Some more scholarly types in the field refer to it as a 'Burch anastigmat', as Burch refers to its existence in his early work on microscope objectives.

The answer was found in Allibone's 42-page monograph on the life of Burch [47]. On page 26, Allibone describes how Burch learned of the existence of the spherical-mirror anastigmat in a letter to Nature by Brumberg [48]. Burch wrote a letter to Nature in response [49], pointing out how interesting Brumberg's solution was and mentioning that he himself was making Schwarzschild aplanats and anastigmats.

Burch demonstrated how this system could be coupled with a Mersenne feeder to produce a telescope (figure 3.26) [45].

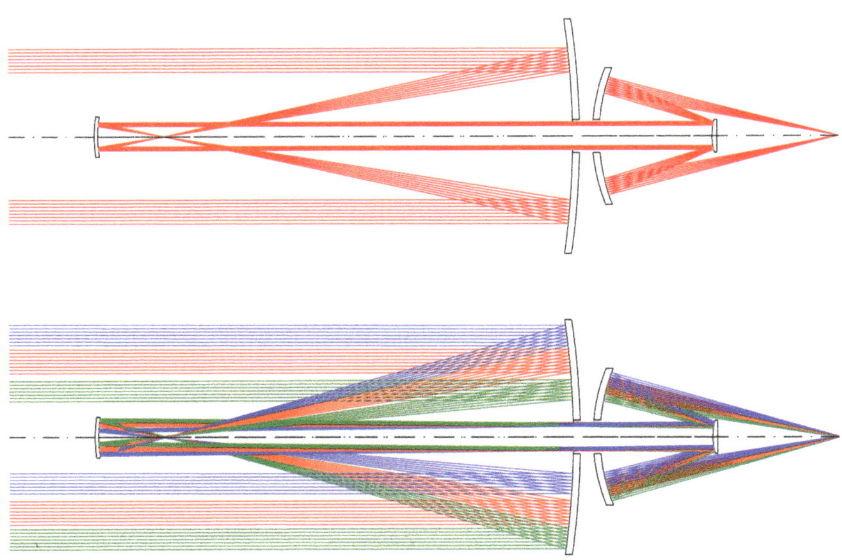

Figure 3.26. Four-mirror anastigmat produced by coupling a Mersenne system (left) with a Brumberg concentric system (right). To achieve a flat field, the Gregorian form of the Mersenne system on the left must be selected, as its field curvature can be made to balance the unavoidable field curvature of the monocentric spheres on the right. Created with Code V.

Given that we have already fully analysed the Mersenne and know for sure that it is an anastigmat, it is a very simple matter to prove that this four-mirror system is anastigmatic.

We know that the mirrors are concentric and that therefore the plates are coincident. We also know that, as M1 is in a collimated light space, both plates are located at the centre of curvature of M1, but this is irrelevant, as the plates must be cancelling if this system is an anastigmat; therefore, they could be anywhere without affecting that correction.

All we have to do to prove that this four-mirror design is indeed an anastigmat is to calculate the spherical aberration of the two spherical mirrors and show that they cancel.

As with the Mersenne, we can pick any combination of dimensions. Given the ratios described above, we can leave r_1, t_1, and r_2 expressed in terms of ϕ, *vis*:

$$r_1 = \phi^2 - \phi,$$

$$t_1 = -\phi,$$

$$r_2 = \phi^2.$$

We can choose any height for our marginal axial ray, y_c, so let us call that $\frac{1}{\phi}$. Using equation (2.29), we have the following for the spherical aberration from M1:

$$W = -\frac{1}{4}nc^3y^4 \rightarrow W_{S1} = -\frac{1}{4} \times 1 \times \frac{1}{(\phi^2 - \phi)^3} \times \frac{1}{\phi^4} = -\frac{1}{4\phi^4}, \tag{3.18}$$

because $(\phi^2 - \phi) = 1$.

To trace the marginal axial ray to M2, we set up a matrix as follows

$$\left(0, \frac{1}{\phi}\right) \cdot \begin{pmatrix} 1 & 0 \\ \frac{-2}{1} & 1 \end{pmatrix} \cdot \begin{pmatrix} 1 & -\frac{-\phi}{-1} \\ 0 & 1 \end{pmatrix} = \left(-\frac{2}{\phi}, \frac{1}{\phi} + 2\right),$$

which, with a little algebra, becomes

$$\left(-\frac{2}{\phi}, \phi^2\right).$$

Next, remembering that $i_2 = \varphi_2 - u_2$ (noting that the lower case $\varphi \neq \phi$), we have:

$$u_1' = \frac{2}{\phi},$$

$$y_2 = \phi^2,$$

$$\varphi_2 = \frac{y_2}{r_2'} = \frac{\phi^2}{\phi^2} = 1, \text{ so}$$

$$i_2 = 1 - \frac{2}{\phi} = -\frac{1}{\phi^3}.$$

We can now calculate the spherical aberration arising from M2 from equation (2.28)

$$W = -\frac{1}{4}nci^2y^2 \rightarrow W_{S2} = -\frac{1}{4} \times -1 \times \frac{1}{\phi^2} \times \left(-\frac{1}{\phi^3}\right)^2 \times \phi^4 = \frac{1}{4\phi^4}. \quad (3.19)$$

So, $W_{S2} = -W_{S1}$, and, as the saying goes, 'we're golden'.

Jokes aside, if any student reading this is becoming at all interested in getting good at ray-tracing calculations, this is not a bad idea for an optical designer. Surely, any commercial ray-tracing code can do it billions of times faster, but for a human, it is a very good mental workout, and for a person who designs optics, it builds awareness in all sorts of ways. This little exercise, where everything is ϕ, serves as an excellent test for any reader who is trying their hand at this, to make sure they understand all the signs, switching refractive indices, directions, etc. as it is more difficult than usual to spot mistakes in this case.

As we leave this chapter on multiple-element plate diagrams, there are a few things to bear in mind. The two-mirror systems had limited geometrical forms. With systems of three mirrors, a much wider range of geometries is possible. This increases as we add more mirrors, but so does the problem of central obscuration in radially symmetrical systems. A solution to this is to use systems that have some combination of the following three elements:

1) Objects that are displaced laterally from the axis, or angular displacement for telescopes.
2) A pupil that is displaced from the axis of symmetry.
3) Generally tilted and de-centred optical elements, that is, systems with no base axial symmetry, which may or may not have bilateral symmetry.

These types of systems are discussed in the following chapters, with type (1) being discussed extensively in the next chapter.

We have seen that for any number of elements, we only require four variables to achieve anastigmatic performance. An advantage, then, of additional elements is that this allows more freedom to choose other desired properties of the system, such as enforcing zero Petzval curvature and suitable first-order geometries, as well as chromatic correction in refracting systems.

A final point before leaving this chapter is to draw the reader's attention to the works of David Shafer. At the time of writing, David has been working as a highly innovative optical designer and a field leader in the design of high-end lithography systems for more than five decades. He has written numerous papers on variants of the monocentric systems and anastigmatic combinations introduced here and often extends to higher orders of aberration correction without resorting to complex surfaces. He has given numerous talks about interesting ways of doing the 'anastigmat algebra' discussed in this chapter. Any student or optics professional who is interested and wants to learn more about the power of monocentric surfaces and the elegant combination of surprisingly simple building blocks that result in complex high-performance systems would do well to look up Shafer's papers [50].

The next chapter deals with a very famous optical system that was patented by David's mentor, Abe Offner, and David has had a lot to do with this and similar systems.

References

[1] Abbe E 1873 Beiträge zur Theorie des Mikroskops und der mikroskopischen Wahrnehmung *Arch. Mikrosk. Anat.* **9** 413–68

[2] Wilson R N 2007 *Reflecting Telescope Optics I: Basic Design Theory and its Historical Development* 2nd edn (Berlin–Heidelberg: Springer)

[3] Chrétien H 1922 Le télescope de Newton et le télescope aplanétique *Revue d'Optique* **1** 49–64

[4] Schwarzschild K 1905 Untersuchungen zur geometrischen Optik II. Theorie der Spiegelsysteme und der astronomischen Instrumente *Math. Ann.* **61** 504–43

[5] Aldis H L 1900 On the construction of photographic objectives *Photogr. J.* **30 June** 291–9

[6] Burch C R 1943 On aspheric anastigmatic systems *Proc. Phys. Soc.* **55** 433–44

[7] Rankine W J M 1858 *A Manual of Applied Mechanics* 2nd edn. (London: Charles Griffin & Co.)

[8] Linfoot E H 1955 *Recent Advances in Optics* (Oxford: Clarendon)

[9] Mersenne M 1644 *Cogitata Physico-Mathematica* (Paris: Marin Mersenne)

[10] Descartes R 1637 La Dioptrique *Discours de la Méthode pour bien conduire sa raison, et chercher la vérité dans les sciences* (Leiden: Jan Maire)

[11] Mersenne M 1636 *De la composition de musique* Harmonie Universelle 4 *(Paris: Sebastien Cramoisy and Pierre Ballard) 60–70*

[12] Gregory J 1663 *Optica Promota: The Advancement of Optics* (London: J. Field)

[13] Cassegrain L 1672 Nouvelle découverte touchant la vision par réflexion, par un moyen de télescope. Recueil des mémoires et conférences concernant les arts et les sciences *J. Sçavans* **4** 255–7

[14] Baker J G 1969 On improving the effectiveness of large telescopes *IEEE Trans. Aerosp. Electron. Syst.* **AES-5** 261–72

[15] Kingslake R and Barry Johnson R 2010 *Lens Design Fundamentals* 2nd edn. (Burlington, MA and Kidlington,: Academic Press/SPIE)

[16] Couder A 1926 Sur un type nouveau de télescope photographique *C. R. Acad. Sci.* **182** 418–20

[17] Vassiliev V V, Fegan S J and Brousseau P 2007 Schwarzschild–Couder two-mirror telescope for ground-based γ-ray astronomy *Astropart. Phys.* **28** 10–27 Also available as arXiv:0708.2741

[18] Schwarzschild K 1905 Investigations on geometrical optics I: introduction to the aberration theory of optical instruments based on the eikonal concept *Abhandlungen der Königlichen Gesellschaft der Wissenschaften zu Göttingen, Mathematisch-Physikalische Klasse* (Neue Folge) **4** 1–31

[19] Bruns H 1895 *Das Eikonal* **21** (Leipzig: Abhandlungen der Königlichen Sächsischen Gesellschaft der Wissenschaften, Mathematisch-Physikalische Klasse) 321–436

[20] Hamilton W R 1833 On a general method of expressing the paths of light, and of the planets, by the coefficients of a characteristic function *Dublin Univ. Rev. Q. Mag.* **1** 795–826

[21] Hamilton W R 1837 Third supplement to an essay on the theory of systems of rays *Trans. R. Irish Acad.* **17** 1–144

[22] Nolte D D 2023 The iconic eikonal and the optical path. *Galileo Unbound* Blog. 'Possibly motivated by his studies done with Hausdorff on refraction of light by the atmosphere, Bruns became interested in Malus' Theorem for the same reasons and with the same goals as Hamilton yet was unaware of Hamilton's work in optics.' https://galileo-unbound.blog/2019/05/30/the-iconic-eikonal-and-the-optical-path/
[23] de Meijere L F and Velzel C H F 1989 Dependence of third- and fifth-order aberration coefficients on the definition of pupil coordinates *J. Opt. Soc. Am.* A **6** 1609–17
[24] Schwarzschild K 1992 Biography of Karl Schwarzschild (1873–1916) *Gesammelte Werke/Collected Works* (Berlin, Heidelberg: Springer) 1–28
[25] Miller J A and Cogshall W A 1940 Optical developments at Indiana University: construction of a 24-inch Schwarzschild reflector in 1928 *Indiana Alumni Magazine*
[26] Burch C R 1947 Reflecting microscope objectives *Proc. Phys. Soc.* **59** 41–57
[27] Dolejší J 1952 New trends in microscopy *Czech. J. Phys.* **2** 35–43
[28] Paul M 1935 Systèmes correcteurs pour réflecteurs astronomiques *Rev. Opt.* **14** 201–4
[29] Dimitroff G Z and Baker J G 1945 *Telescopes and Accessories* (Philadelphia: Blakiston)
[30] Rakich A 2000 A complete survey of three-mirror anastigmatic reflecting telescope systems with one aspheric surface *MSc thesis* (University of Canterbury, Christchurch, New Zealand)
[31] Feinberg L, Menzel M, Van Campen J, Parrish K, Geithner P, McElwain M, Bolcar M and Sitarski B 2025 James Webb Space Telescope: Lessons Learned and Applications to Future Flagships (Bellingham, WA: SPIE) 124480L
[32] Tamai R, Koehler B, Cirasuolo M, Biancat-Marchet F, Tuti M, González-Herrera J-C and Ramsay S 2024 ESO's ELT halfway through construction *Proc. of SPIE 13094: Ground-based and Airborne Telescopes X* (SPIE) p 1309415
[33] Krabbendam V L and Thomas S 2024 The Vera C. Rubin construction status in 2024 *Proc. of SPIE 13094: Ground-based and Airborne Telescopes X* (SPIE) p 130940J
[34] Gilmore A C 2007 Norman J. Rumsey (15 September 1922–9 January 2008) *N. Z. Sci. Mon.* (issue dated 19 January 2007)
[35] Rumsey N J 1969 A compact three-reflection astronomical camera *Optical Instruments and Techniques* (Newcastle upon Tyne: Oriel Press) 514–20
[36] Krabbendam V L 2008 The large synoptic survey telescope concept design overview *Proc. of SPIE 7012: Ground-based and Airborne Telescopes II* (SPIE) p 701205
[37] Lemaître G R 1984 Optical Design with the Schmidt Concept 1. Ground-Based Development 2. The Space Schmidt Project for the 1990's? *Astronomy with Schmidt-Type Telescopes* **vol 110** ed M Capaccioli (Dordrecht: Reidel) 533–48 IAU Colloq. 78, Asiago
[38] Lemaitre G R 1979 Optique astronomique—sur la résolution des télescopes de Schmidt de type catoptrique *C. R. Acad. Sci., Paris* **288B** 297–9
[39] Su D-Q, Cui X, Wang Y and Yao Z 1998 Large-sky-area multi-object fiber spectroscopic telescope (LAMOST) and its key technology *Proc. SPIE* **3352** 76–90
[40] Lemaître G R, Montiel P, Joulié P, Dohlen K and Lanzoni P 2005 Active optics and modified-Rumsey wide-field telescopes: MINITRUST demonstrators with vase- and tulip-form mirrors *Appl. Opt.* **44** 7322–32
[41] Willstrop R V 1984 The Mersenne–Schmidt: a three-mirror survey telescope *Mon. Not. R. Astron. Soc.* **210** 597–609
[42] Rakich A and Rumsey N J 2002 Method for deriving the complete solution set for three-mirror anastigmatic telescopes with two spherical mirrors *J. Opt. Soc. Am.* A **19** 1398–405

[43] Rumsey N J 1969 Telescopic system utilizing three axially aligned substantially hyperbolic mirrors *U.S. Patent* 3460886, filed 17 June 1966, granted
[44] Lemaître G R 2009 *Astronomical Optics and Elasticity Theory: Active Optics Methods Astronomy and Astrophysics Library* (Berlin, Heidelberg: Springer) 235–7
[45] Burch C R 1979 Application of the Plate Diagram to reflecting telescope design *Opt. Acta* **26** 493–504
[46] Willard B C 1993 The golden ratio in optics *Opt. Photonics News* **4** 22–5
[47] Allibone T E 1984 Cecil Reginald Burch, 12 May 1901—19 July 1983 *Biogr. Mem. Fellows R. Soc.* **30** 2–42
[48] Brumberg E M 1943 Reflecting microscopes involving two spherical mirrors *Nature* **152** 357
[49] Burch C R 1943 Reflecting microscopes *Nature* **152** 748
[50] Shafer D R 2005 Some odd and interesting monocentric designs *Proc. SPIE 5865: Tribute to Warren J. Smith: A Legacy in Lens Design and Optical Engineering* (SPIE) 586508

Chapter 4

Finite-conjugate systems

In 1973, US Patent No. 3748015, 'Unit Power Imaging Anastigmat' was taken out by Abe Offner and assigned to Perkin-Elmer Corp [1]. It describes a remarkable reflecting optical system. The system is corrected for all third-order aberrations, including Petzval curvature, and all fifth-order aberrations, with the exception of fifth-order astigmatism. It achieved this high degree of correction with only three spherical mirrors. From the previous chapter's results on balanced moments, we can immediately deduce that the mirrors in this system must be concentric, as the resultant coincident plates are the only possible way for three plates to produce an anastigmat. Plate-diagram consciousness! This chapter describes the Offner design, along with a design by Reed [2] that Offner cites as prior art in his patent, and some interesting insights and extensions into this design that came about via plate-diagram analysis.

4.1 The Offner/Reed monocentric relay

Offner's patent cites Reed's US Patent No. 3190171 as prior art (see reference [2]), so we will begin with that. Figure 4.1 shows the system patented by Reed.

Reed's system is described as a useful system for use in flight simulators. Light from distant objects entering the system at the entrance pupil (top left) is relayed to the pupil of a pilot's eye (bottom left) at unit magnification. The invention is described as comprising three spherical reflective surfaces with a common centre of curvature. The two concave mirrors M1 and M3 have twice the radius of curvature of M2, so M1 and M3 form parts of a continuous spherical surface. The entrance pupil and exit pupil lie in a plane that contains the centre of curvature of the mirrors, and the system reference axis is perpendicular to this plane, passing through the common mirror centres.

Reed's patent makes no mention of aberration and describes the system from first-order considerations only. If we investigate the aberrations, we see it suffers from astigmatism, which occurs at a similar transverse dimension to that of the

doi:10.1088/978-0-7503-3099-2ch4

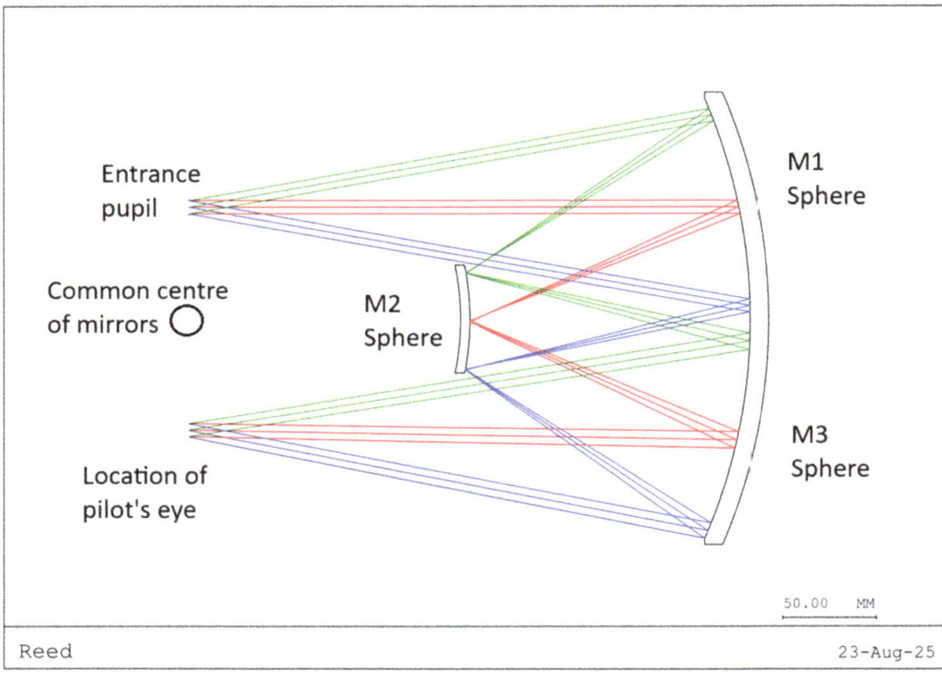

Figure 4.1. Reed's pupil relay with three concentric spherical mirrors. Created with Code V.

pilot's eye pupil, so the system as described would not work very well for its intended purpose (figures 4.2 and 4.3).

The geometry in figure 4.1 means that only an off-axis circular region of the spherically aberrated wavefront arising at each concave mirror is used. As shown in chapter 2 (see figures 2.8 and 2.9 and equation (2.30)), off-axis circular regions of a spherically aberrated wavefront give rise to asymmetrical aberrations. Another way of looking at this is that we see the aberration of tilted spheres here, and astigmatism dominates.

The amount of pupil aberration in Reed's system would have made it quite impractical for its stated purpose.

Coming to Offner's patent, we see the geometry of Reed, but Offner is now considering the system not as a pupil relay but as an imaging system, with a stated application in lithography. Offner clearly describes the aberration correction for this optical conjugation, not only for Reed's geometry but for a much more general class of at least three mirrors. The more general classes are discussed in the following section, but for now, we will stick with Reed's geometry, where M1 and M3 are of equal radius and M2 has half that radius.

Offner points out that this geometry is corrected for field curvature of all orders (as is clear from equation (2.33) applied to the case where $r_1 = r_3 = -2r_2$). Also, spherical aberration, coma, and distortion are corrected to the 5th order, and astigmatism is corrected to the 3rd order. A residual of 5th-order astigmatism can be

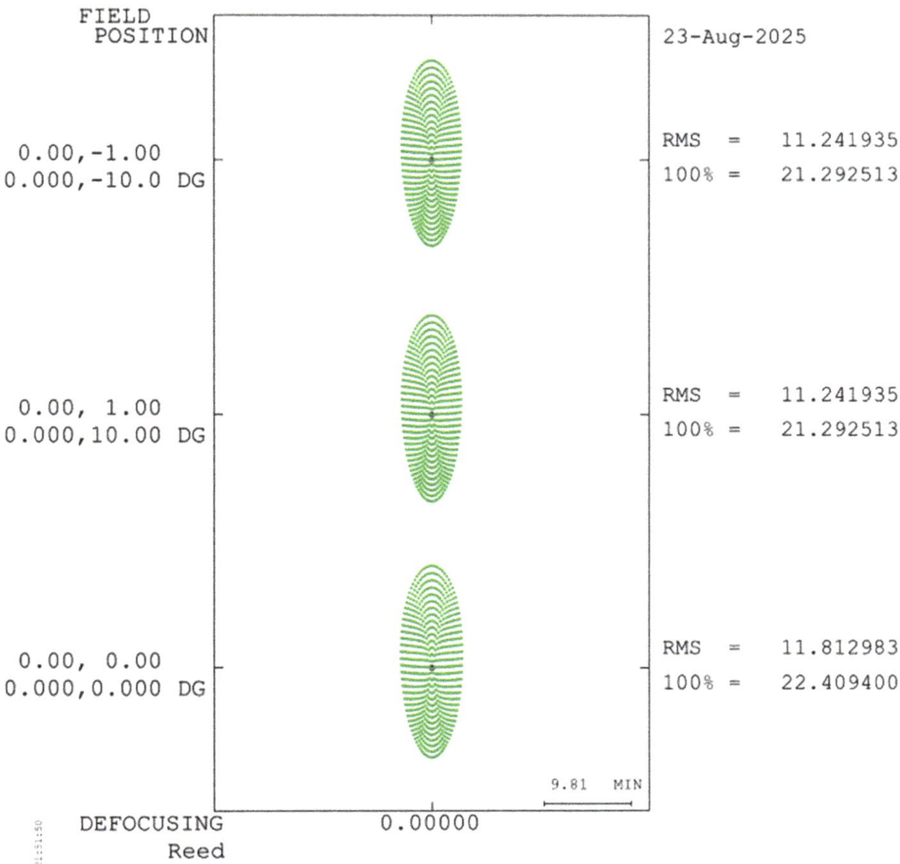

Figure 4.2. Approximately 7 milliradians, or 23 arcminutes, of transverse astigmatism at the pupil would mean that the images formed by the pilot's eye would be blurred by more than a factor of 20 beyond the human eye's resolution limit. Created with Code V.

corrected by a small respacing, reintroducing a balancing 3rd-order astigmatism. Figure 4.4 shows the reconjugated Reed system as presented by Offner, and figures 4.5–4.6 illustrate the aberration correction.

The 5th-order astigmatism can be balanced by perturbing the system in a way that reintroduces a balancing 3rd-order astigmatism (figure 4.7), with the resultant aberration correction shown in figures 4.8 and 4.9.

There are two ways in which Offner systems are commonly perturbed. One is to leave the system rotationally symmetric and monocentric but break the Petzval correction, i.e. M2 has a different radius from half the average of M2 and M3, and the airspace is adjusted to maintain concentricity. Because it is still monocentric and object-centred, spherical aberration, coma, and distortion remain corrected, leaving only astigmatism. The amount of 3rd-order astigmatism can be chosen to balance the 5th-order astigmatism at the centre of the field, i.e. the middle radius of the ring field. Unfortunately, the astigmatism grows linearly with distance from the centre of

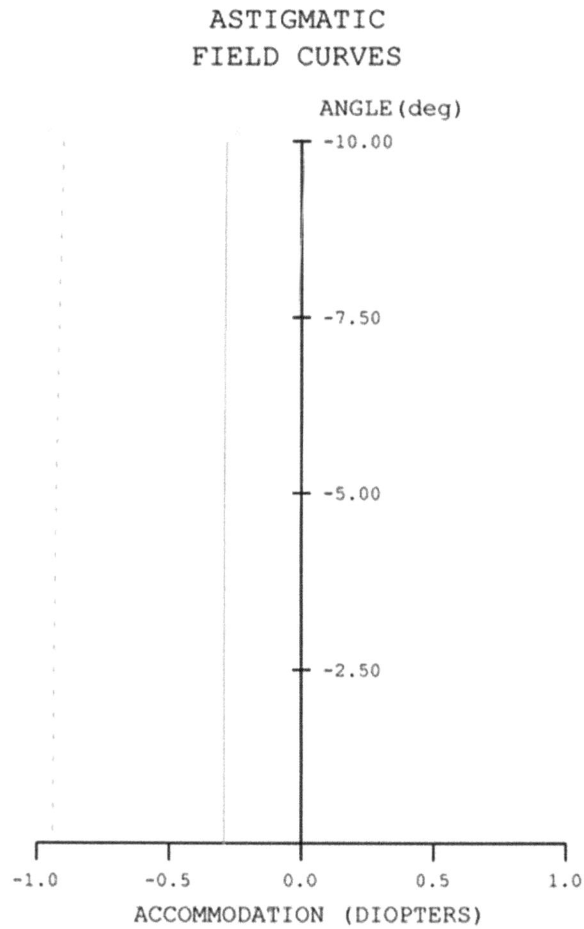

Figure 4.3. The astigmatism of the pupil image is constant with the field. This astigmatism arises because the spherical mirrors are being used off axis. Created with Code V.

the ring, so the usable width of the ring in the radial direction is quite small. Nevertheless, this kind of design was used as a ring-field scanner in the early days of photolithography.

This example shows the 'second level' of correction, wherein the *S* and *T* curves are tangent to each other at the centre of the field, so the residual astigmatism grows quadratically rather than linearly with distance from the centre of the ring; for that reason, the useful field is larger.

One of the things done to achieve this 'tangential correction' was to tilt M1, and this is useful for an application that involves a small rectangular field. For ring-field lithography, this approach is not commonly used, because it would adversely affect the correction with rotation around the ring, i.e. in and out of the page. This shortens the useful transverse width of the ring when used as a push broom scanner.

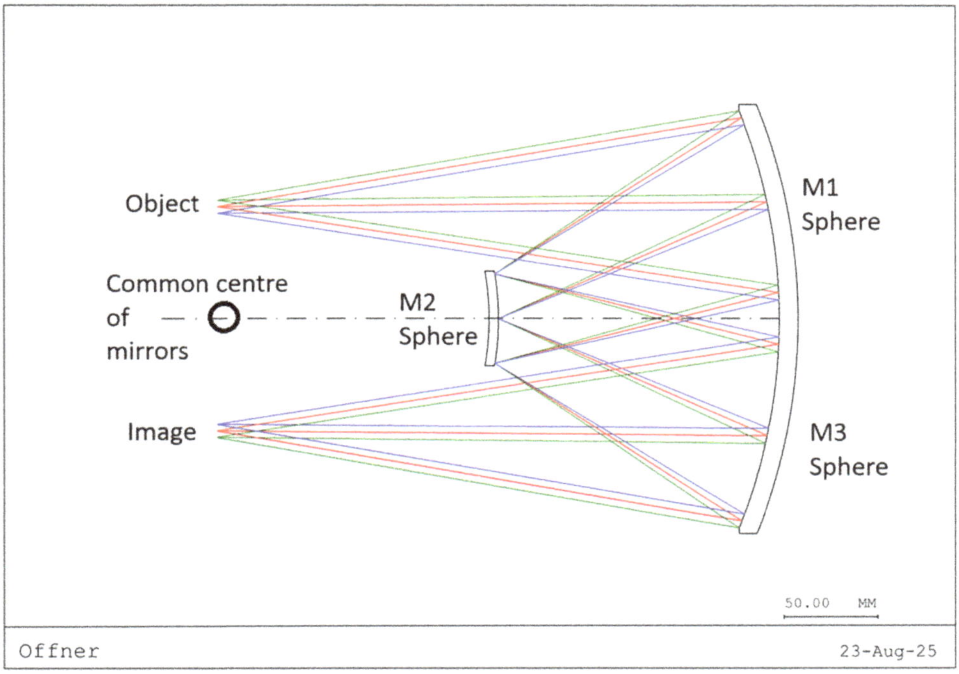

Figure 4.4. Comparing this figure to figure 4.4, we see an identical geometrical layout. The only difference is that in the Reed system, the object was nominally at infinity (or at least distant), and the optics relayed the pupil. In the Offner system, the object and pupil conjugates are reversed, and now the pupil is at infinity, so the system has parallel input principal rays, i.e. it is telecentric. Note that the telecentric principal rays are focused by M1 at the principal focus, which is located at and has the same curvature as M2. Created with Code V.

In figure 4.9, we can see that the combination of third- and fifth-order astigmatism thus obtained makes a turning point in the tangential astigmatism curve centred on the 60 mm off-axis midpoint of the ring field.

The combination of remarkable relative simplicity and excellent correction has led to the Offner becoming a standard unit-magnification relay over the 50+ years since its invention. In addition to its original application in lithography, it has found an array of applications, from industrial surface inspection systems to high-end astronomical instrumentation to space-borne spectrographs. This latter application, the ‘Offner spectrograph’, was developed by the Jet Propulsion Laboratory (JPL). Wanting to take advantage of the beneficial features of the Offner as related here, JPL made a large investment in developing a convex curved reflective diffraction grating that could be placed at the M2 location in the Offner design. Given its ubiquitous application, most students of optics are taught about it as part of their coursework.

Considering the plate diagram of the Offner system, two things stand out. Firstly, concentric mirrors mean coincident plates. Three plates lie on top of each other at the image of the mutual centre of curvature of the mirrors in a collimated light space. The Offner system has no collimated light space, and as it is object centred and

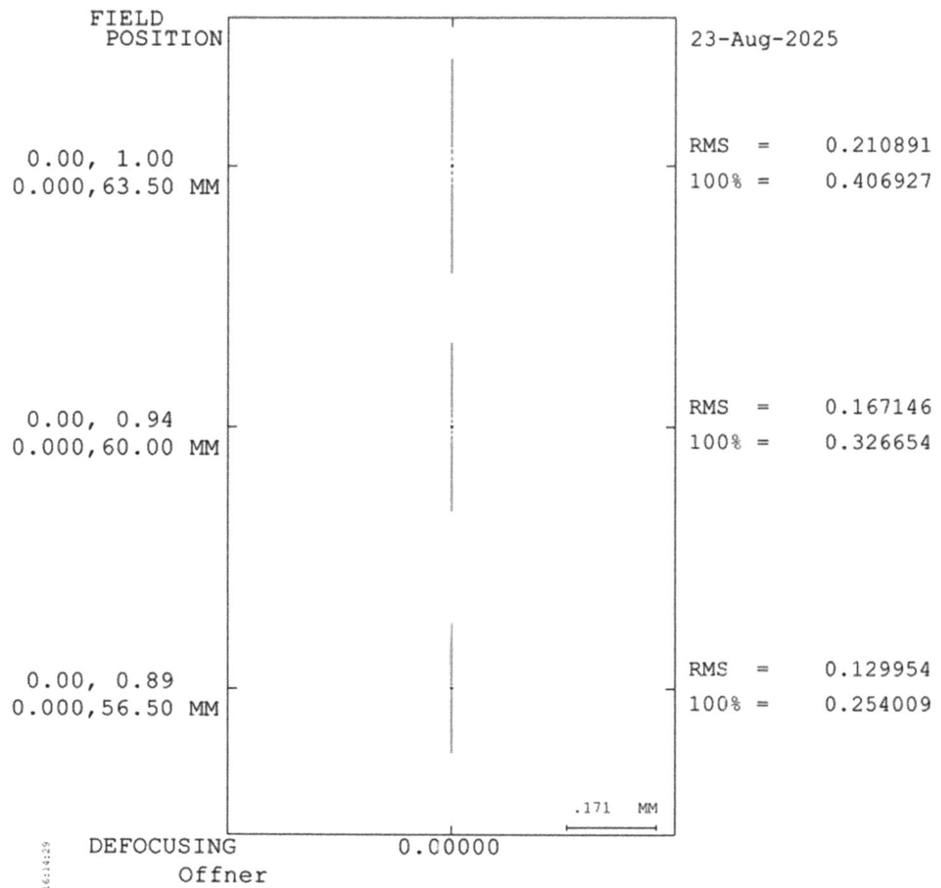

Figure 4.5. The spot diagrams show a line significantly larger than the Airy disc at this numerical aperture (0.17) and field. Created with Code V.

telecentric, this means that when we collimate the light from the Offner to make a space for the plate diagram, the pupil and object conjugates are swapped: the pupil goes to a conjugate of the Offner object surface, and the centres of curvature of the mirrors are imaged at infinity.

Also, as the object surface is in a plane containing the mirror centres, being the natural conjugation for a sphere, there is no spherical aberration from any of the mirrors. The spherical aberration is calculated from the axial marginal ray. For an object-centred sphere, the axial marginal ray is a radius and strikes the mirror with a zero angle of incidence in all cases.

This means each individual plate of the Offner is of zero 'strength'. The essential result is that we have, for the plate diagram of the Offner, three flat windows located at an infinite distance from a pivot point at the object plane of the Offner. Clearly, we should expect zeroed moments in such a degenerate case.

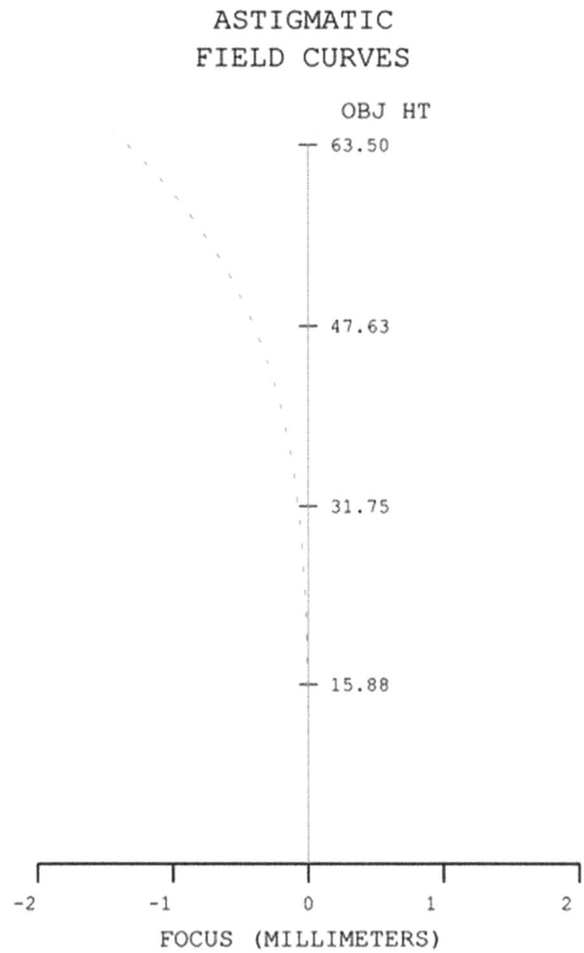

Figure 4.6. The field curvature plot shows that for fields above 20 mm, the sagittal focal surface remains flat but the tangential focal surface is curved by 5th-order astigmatism. Created with Code V.

It would seem at first as if that is that, and the analysis is complete; there is no need to investigate further. The lack of aberration is fully explained. However, there is a very surprising subtlety relating astigmatism to Petzval curvature, which has opened up new levels of understanding of the Offner system, new design insights, and some discoveries related to basic aberration theory.

4.2 A generalised Offner design

In 2017, the author presented a paper at the International Optical Design Conference (IODC) in Denver, surveying TMA optical design in the 20th century [3]. At one point, when discussing the Offner, he made a mistake. In a brief comment on the Offner system, it was stated:

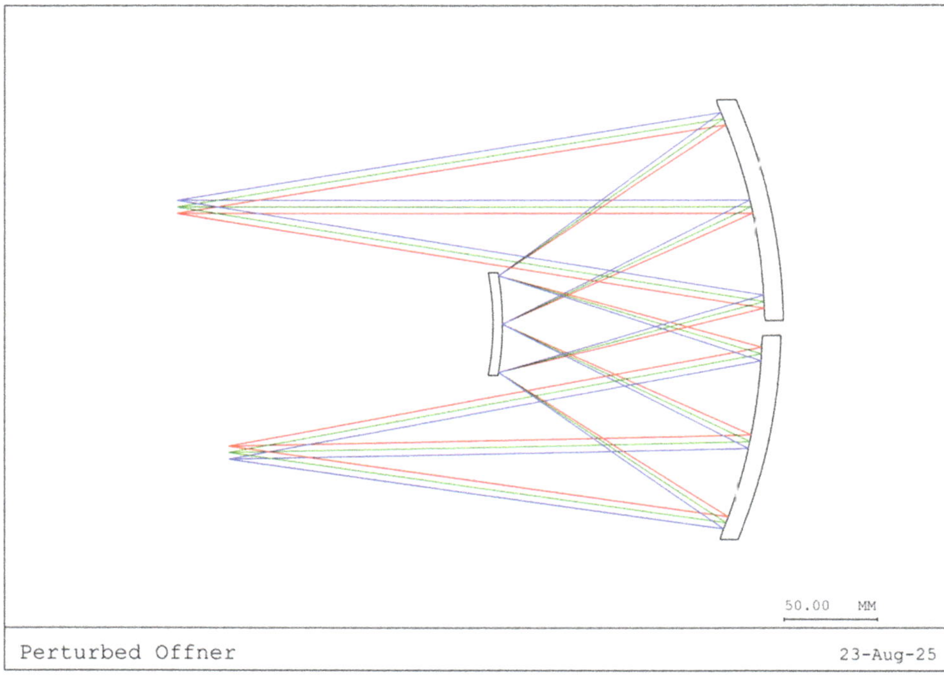

Figure 4.7. Modified Offner where small perturbations are made to M1's position, spacings, and radii, allowing a precise balancing of 3rd- and 5th-order astigmatism at the ring field with a radius of 60 mm. The pupil still forms at M2. Created with Code V.

> *...and the three plates, coincident at the common centre of curvature, cancel.*

As was just discussed, the plates do not lie at the centre of curvature, but rather at infinity. In 2012, while working for the European Southern Observatory on some concepts for the Mid-Infrared E-ELT Imager and Spectrograph (METIS) [4] and High Angular Resolution Monolithic Optical and Near-infrared Integral (HARMONI) [5] optical relays, two instruments for the upcoming Extremely Large Telescope (ELT), the author had found via ray tracing that the Offner could be modified to provide a pupil that lay in free space, that was not on the surface of M2, and that was clear of the volume containing rays to and from M2.

This 'broken-symmetry' Offner was useful for locating a cold stop. At least one other designer, Tibor Agócs, has independently found the same result [6] by ray tracing [7, 8], but there have been no theoretical studies of the aberrations and first-order properties of such systems.

In 2019, realising the mistake about plate location and the significance of variants of the Offner with clear pupils, the author wondered how to conduct a plate-diagram analysis for this degenerate case. While discussing this (private communication, 2019), Dr John Rogers suggested that by adding a small Δ to the axial location of

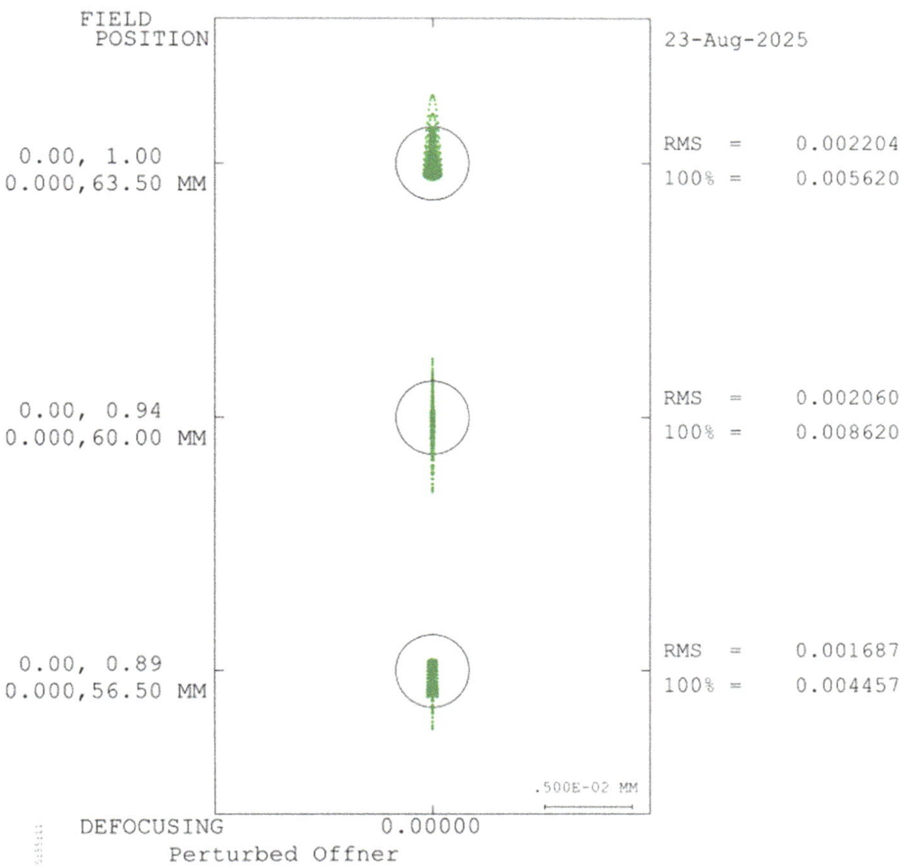

Figure 4.8. Balanced astigmatism leads to diffraction-limited performance, with spot blurs reduced in diameter by two orders of magnitude in this case. Created with Code V.

the object, expressions could be found that could then be evaluated in the limit where Δ goes to zero.

The upshot was two papers: one used the plate diagram [9], while the other used a Seidel sum analysis [10] to investigate the properties of this system. Both papers, read together, provide a very interesting insight into how looking at the same problem in different ways can usefully broaden understanding. These papers showed that all of the properties of the symmetrical Offner system apply to a much broader class of monocentric, object-centred spherical systems. The key is that all fully corrected monocentric systems require Petzval curvature to be corrected.

The following derivation replicates the results of the Seidel sum analysis of reference [10] but uses the plate diagram to get there.

We can begin by calculating the plate strengths W_i for each of the three plates when a small axial delta is introduced into the object location, using quantities as indicated in figure 4.10.

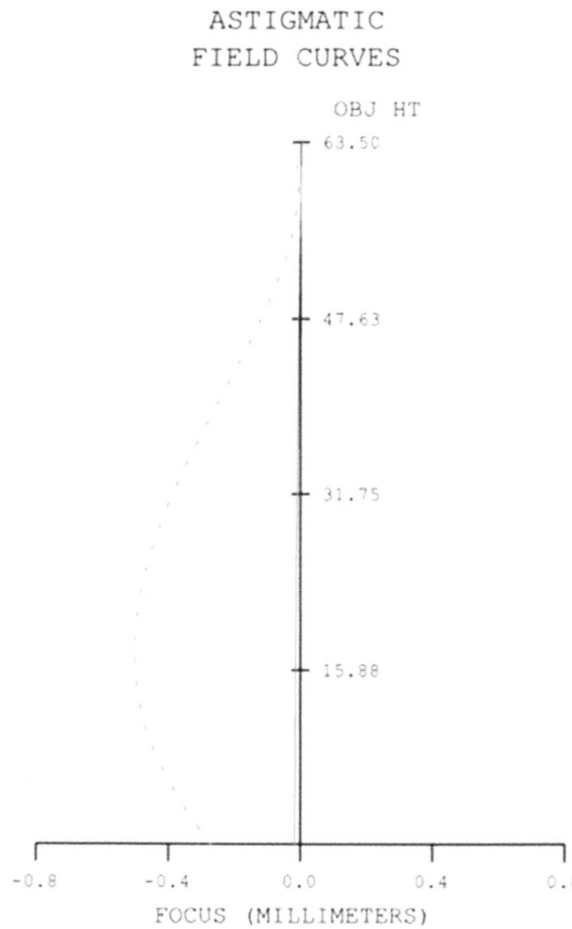

Figure 4.9. In this figure, we see that 3rd-order astigmatism is balanced with 5th-order astigmatism at a height of 60 mm, which is the centre of the field. Created with Code V.

We are interested in information about the general combination of three mirrors that are monocentric, nominally object centred with an axial perturbance Δ, and have three different radii. We define our variables to match this general condition:

$$t_1 = -(r_1 + \Delta), \;\; t_2 = r_1 - r_2, \;\; t_3 = -(r_3 - r_2), \;\; t_4 = r_3 - \Delta, \tag{4.1}$$

remembering that our t_i have both magnitude and direction. As light initially travels from left to right in air, the refractive index in the initial space, n_0, is positive 1, and the sign alternates on reflection. The effect of alternating sign, unity magnitude refractive index and alternating sign directions of light travel is that the quantity $\frac{t_i}{n_i}$ is sign invariant, always negative in this case.

Our initial launch vector is:

$$(n_0 u_0, \; y_0) = (u_0, \; 0),$$

for the marginal ray of the axial pencil of rays.

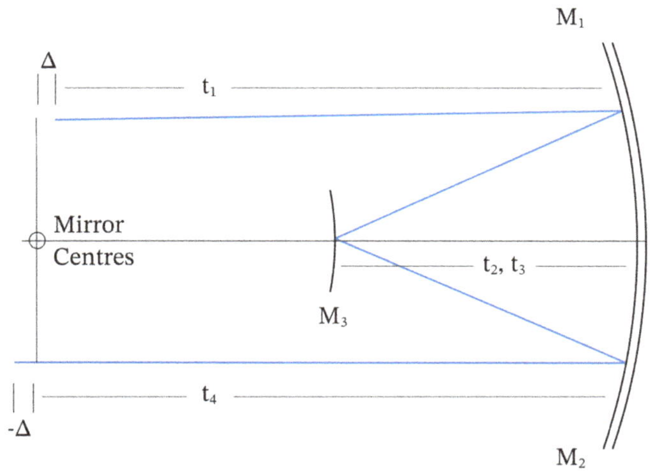

Figure 4.10. Quantities used in the Offner plate diagram. The system here is a ‘classical Offner’ with M1 and M3 radii equal and M2 of half the radius of M1, but the derivation given here applies to any combination of the three mirror radii.

Our matrices can therefore be defined as:

$$A = \begin{pmatrix} 1 & -\frac{t_1}{n_0} \\ 0 & 1 \end{pmatrix} = \begin{pmatrix} 1 & -\frac{-(r_1+\Delta)}{1} \\ 0 & 1 \end{pmatrix} = \begin{pmatrix} 1 & r_1 + \Delta \\ 0 & 1 \end{pmatrix},$$

$$B = \begin{pmatrix} 1 & 0 \\ \frac{n_1 - n_0}{r_1} & 1 \end{pmatrix} = \begin{pmatrix} 1 & 0 \\ \frac{-2}{r_1} & 1 \end{pmatrix},$$

$$C = \begin{pmatrix} 1 & -\frac{t_2}{n_1} \\ 0 & 1 \end{pmatrix} = \begin{pmatrix} 1 & -\frac{r_1 - r_2}{n_1} \\ 0 & 1 \end{pmatrix} = \begin{pmatrix} 1 & r_1 - r_2 \\ 0 & 1 \end{pmatrix},$$

$$D = \begin{pmatrix} 1 & 0 \\ \frac{n_2 - n_1}{r_2} & 1 \end{pmatrix} = \begin{pmatrix} 1 & 0 \\ \frac{2}{r_2} & 1 \end{pmatrix},$$

$$E = \begin{pmatrix} 1 & -\frac{t_3}{n_2} \\ 0 & 1 \end{pmatrix} = \begin{pmatrix} 1 & -\frac{-(r_3 - r_2)}{n_2} \\ 0 & 1 \end{pmatrix} = \begin{pmatrix} 1 & r_3 - r_2 \\ 0 & 1 \end{pmatrix}. \tag{4.2}$$

Of course, the production of these matrices can be scripted, but if doing this manually, it is good practice to write them out by hand this way, because it is easy to make a sign error, and one sign error can result in a wasted hour.

We wish to produce the quantities i_j and y_j at the jth surface to evaluate the three W_j.

We begin by reflecting a marginal axial ray from M1. Using the scheme given in equation (2.13), we produce:

$$(n_0u_0,\ 0).\ A = (u_0,\ 0).\begin{pmatrix} 1 & r_1 + \Delta \\ 0 & 1 \end{pmatrix} = (u_0,\ (r_1 + \Delta)u_0), \tag{4.3}$$

from which,

$$y_1' = (r_1 + \Delta)u_0, \tag{4.4}$$

and u_0, which has experienced no reflection, has not changed. No change in u is expected here because spherical aberration is evaluated with the incident i:

$$\varphi_1 = \frac{y_1}{r_1} = \frac{(r_1 + \Delta)u_0}{r_1}.$$

$$i_1 = \varphi_1 - u_0 = u_0\left(\frac{(r_1 + \Delta)}{r_1} - 1\right) = \frac{\Delta \times u_0}{r_1}. \tag{4.5}$$

$$\begin{aligned} W_1 &= -\frac{1}{4}nci^2y^2, \\ &= -\frac{1}{4}n_0\frac{1}{r_1}\left(\frac{(\Delta \times u_0)}{r_1}\right)^2((r_1 + \Delta)u_0)^2, \\ &= -\frac{u_0^4\Delta^2}{4r_1^3}(r_1 + \Delta)^2, \\ &= -\frac{u_0^4}{4}\left(\frac{\Delta^4}{r_1^3} + \frac{2\Delta^3}{r_1^2} + \frac{\Delta^2}{r_1}\right). \end{aligned} \tag{4.6}$$

From this, we see that the object-conjugate-shift dependence of spherical aberration for a single spherical mirror, where the shift is measured from the natural conjugate of a sphere at its centre, has quartic, cubic, and quadratic dependence on Δ.

We can do a quick sanity check here. As Δ tends to infinity, only the quartic dependence is significant. Equation (4.2) shows that as $\Delta \rightarrow$ infinity, $\Delta \times u_0 \rightarrow y_1'$, and $-\frac{u_0^4}{4}\left(\frac{\Delta^4}{r_1^3} + \frac{2\Delta^3}{r_1^2} + \frac{\Delta^2}{r_1}\right) \rightarrow -\frac{y_1^4}{4r_1^3}$, as it should, according to our equation (2.29).

We now continue the ray trace to M2, using matrices B and C together with our initial trace:

$$(n_0u_0,\ 0).\ A.\ B.\ C = \left(-u_0\left(1 + \frac{2\Delta}{r_1}\right),\ u_0\left(r_2 + \Delta\left(\frac{2r_2}{r_1} - 1\right)\right)\right). \tag{4.7}$$

We have therefore the following at M2, remembering that the angle term in equation (4.3) is divided by $n_1 = -1$:

$$y_{1+}' = y_2 = u_0(r_2 + \Delta\left(\frac{2r_2}{r_1} - 1\right),$$

and

$$u_1 = u_0\left(1 + \frac{2\Delta}{r_1}\right). \tag{4.8}$$

Continuing,

$$\varphi_2 = \frac{y_2}{r_2} = u_0(\frac{2\Delta}{r_1} - \frac{\Delta}{r_2} + 1),$$

$$i_2 = \varphi_2 - u_1 = -\frac{\Delta \times u_0}{r_2}. \tag{4.9}$$

With these, we can evaluate W_2:

$$\begin{aligned} W_2 &= -\tfrac{1}{4}nci^2y^2, \\ &= -\tfrac{1}{4}n_1\tfrac{1}{r_2}\left(\tfrac{(-\Delta \times u_0)}{r_2}\right)^2\left(u_0\left(r_2 + \Delta\left(\tfrac{2r_2}{r_1} - 1\right)\right)\right)^2, \\ &= -\frac{u_0^4\Delta^2}{4r_1^2r_2^3}(2\Delta r_2 - \Delta r_1 + r_1r_2)^2, \end{aligned}$$

and expanding this gives

$$W_2 = \frac{u_0^4\Delta^4}{4}\left(\frac{1}{r_2^3} - \frac{4}{r_2^2r_1} + \frac{1}{r_2^2r_1}\right) + \frac{u_0^4\Delta^3}{4}\left(\frac{4}{r_2r_1} - \frac{2}{r_2^2}\right) + \frac{u_0^4\Delta^2}{4}\left(\frac{1}{r_2}\right). \tag{4.10}$$

Again, we note that W_2 goes to zero as Δ goes to zero, as it should.

Continuing in the same fashion to M3, we evaluate

$$(n_0u_0, 0).\ A.\ B.\ C.\ D.\ E.$$

Going through the same steps as above to evaluate W_3 and grouping terms as in equation (4.10), we get

$$\begin{aligned} W_3 = &\frac{u_0^4\Delta^4}{4}\left(\frac{-1}{r_3^3} + \frac{4}{r_3^2r_2} - \frac{4}{r_3^2r_1} - \frac{4}{r_3r_2^2} - \frac{4}{r_3r_1^2} + \frac{8}{r_1r_2r_3}\right) \\ &+ \frac{u_0^4\Delta^3}{4}\left(\frac{-2}{r_3^2} + \frac{4}{r_3r_2} - \frac{4}{r_3r_1}\right) \\ &- \frac{u_0^4\Delta^2}{4}\frac{1}{r_3^3}. \end{aligned} \tag{4.11}$$

Summing the W_i expanded in this way gives us an expression for the spherical aberration of three monocentric spheres of any radius with an axial offset of Δ from the plane containing the common centres of curvature. This expression contains 19 distinct terms. Burch's miasma is thick around us at this point, but things will clear up soon.

Recalling the Petzval condition from equation (2.33) and applying it to our three mirrors, we have:

$$-\frac{2}{r_1} + \frac{2}{r_2} - \frac{2}{r_3} = C_p. \tag{4.12}$$

We can divide the system spherical aberration sum, $\sum_{i=1}^{3} W_i = W_{SYS}$, into three subsets, depending on their power dependency on Δ. With some rearrangement (and the Mathematica software is very useful for this step), we can arrive at the following.

In terms dependent on Δ^4, Δ^3, and Δ^2, W_{SYS} can be expressed as:

$$W_{SYS\Delta^4} = -\frac{u_0^4\Delta^4}{4}\left(\frac{r_1^3 r_2(r_2 - 2r_3)^2 + 4r_2^3 r_1 r_3(r_1 + r_3) + r_1^3(r_2 - r_1)(r_1^2 - 3r_1 r_2 + r_2^2)}{r_1 r_2 r_3}\right). \tag{4.13}$$

$$\begin{aligned} W_{SYS\Delta^3} &= -\frac{u_0^4\Delta^3}{2}\left(\frac{(r_1(r_2 - 2r_3) + r_2 r_3)^2}{r_1 r_2 r_3}\right),\\ &= \frac{u_0^4\Delta^3}{4}\left(-\frac{2}{r_1} + \frac{2}{r_2} - \frac{2}{r_3}\right)\left(\frac{1}{r_1} - \frac{1}{r_2} + \frac{1}{r_3}\right),\\ &= -\frac{u_0^4\Delta^3}{2}C_P^2. \end{aligned} \tag{4.14}$$

$$\begin{aligned} W_{SYS\Delta^2} &= \frac{u_0^4\Delta^2}{4}\left(-\frac{1}{r_1} + \frac{1}{r_2} - \frac{1}{r_3}\right),\\ &= \frac{u_0^4\Delta^2}{8}C_P. \end{aligned} \tag{4.15}$$

To summarise, $W_{SYS}(\Delta)$ can be broken into three parts. The first part, dependent on Δ^4, is not strictly dependent on Petzval curvature, but the second part, which is dependent on Δ^3, is at the same time dependent on C_P^2, while the third part, which is dependent on Δ^2, is at the same time linearly dependent on C_P.

Already, we see that the Petzval sum is a strong differentiator in the aberration characteristics of sets of monocentric spheres. To drive Petzval curvature to zero, for any combination of r_1 and r_2, we must have:

$$r_3 = \frac{r_1 r_2}{r_1 - r_2}. \tag{4.16}$$

If we apply this condition, the two parts of $W_{SYS}(\Delta)$ that have Petzval curvature dependency vanish, and equation (4.13) reduces to:

$$\begin{aligned} W_{SYS}(\Delta) &= -\frac{u_0^4\Delta^4}{4}\frac{(r_1 - r_2)}{r_1^2 r_2^2}\\ &= -\frac{u_0^4\Delta^4}{4}\frac{1}{r_1 r_2 r_3}. \end{aligned} \tag{4.17}$$

We now see that the 'fog' has mostly cleared.

It is a significant point to realise that Petzval-corrected monocentric systems are much less sensitive to aberration from small object-conjugate shifts than are systems with residual Petzval curvature.

To complete the plate-diagram analysis for coma and astigmatism, we must consider the common value of x, the distance of the co-located plates from the pupil, in some artificially introduced collimated light space. This part of the analysis is what necessitated the use of Δ in the first place: if Δ is exactly zero, the plates are at infinity, and the system breaks down.

In fact, for the purpose of this analysis, it is not necessary to calculate a value for x at all, as we are merely interested in the behaviour of aberrations as Δ goes to zero.

It suffices to say that

$$x = \frac{A}{\Delta},$$

where A is some constant value. Using this, we can investigate system coma and astigmatism.

From the plate equations, we have

$$\text{coma}_{SYS} \propto \frac{A}{\Delta} W_{SYS},$$

$$\text{astigmatism}_{SYS} \propto \frac{A^2}{\Delta^2} W_{SYS}. \tag{4.18}$$

We can simplify the Δ^4-dependent term in equation (4.13), as we are not currently interested in the details of any terms inside parentheses; therefore, replacing these with $f(r_i)$ and combining the components of W_{SYS} from equations (4.13), (4.14), and (4.15) into one expression, we get:

$$W_{SYS} = \frac{u_0^4 \Delta^2}{2}\left(\frac{C_P}{4} - \Delta C_P^2 - \frac{\Delta^2}{2}(f(r_i))\right). \tag{4.19}$$

Finally, substituting this expression into equation (4.18), we see that

$$\text{coma}_{SYS} \propto A\frac{u_0^4 \Delta}{2}\left(\frac{C_P}{4} - \Delta C_P^2 - \frac{\Delta^2}{2}(f(r_i))\right), \tag{4.20}$$

$$\text{astigmatism}_{SYS} \propto A^2\frac{u_0^4}{2}\left(\frac{C_P}{4} - \Delta C_P^2 - \frac{\Delta^2}{2}(f(r_i))\right). \tag{4.21}$$

This is a remarkable collection of results. The spherical aberration in equation (4.20) evaluates to zero when Δ is zero, of course. But in the presence of Petzval curvature, equation (4.19) shows additional aberration that remains significant for magnitudes of Δ of the order of $\sqrt{C_P}$.

For a single spherical mirror, the term that is quadratic in Δ dominates, as we see in equation (4.6).

Equation (4.19) shows us that in the near region of focus, the spherical aberration of a system of three monocentric spheres grows from zero significantly more quickly with conjugate shift away from the common centres when the system has a finite Petzval sum than when this sum is zero. In the Petzval-corrected triplet, the quadratic and cubic dependencies on Δ are eliminated. For small errors in perfect alignment, as would be seen in a real lithography system, this is something to consider. For large conjugate shifts, the quartic term dominates.

Equations (4.20) and (4.21) reveal new features in our understanding of systems of object-centred monocentric spheres. First, when Δ is zero, coma goes to zero as well in all cases. When we look at the astigmatism expression, we see something different. In this case, there is a term that remains finite as Δ becomes zero. From this, we can say that a system of object-centred monocentric spheres has astigmatism that is linearly dependent on its Petzval sum.

Finally, as the derivation was, in general, for three spheres of any radius, these results show that iff the Petzval sum is zero, ANY combination of three object-centred spheres is an anastigmat.

This is an interesting reminder of the subtleties of indeterminate forms: in this example, when $\Delta \to 0$, $x \to \infty$, and $W_{SYS} \to 0$, equations (4.20) and (4.21) show that a term proportional to $0 \times \infty$ tends to zero, whereas a term proportional to $0 \times \infty^2$ can be finite.

Besides understanding the third-order properties, it is also necessary to consider the first-order properties of such a system of spheres. Do the asymmetric Offner variants maintain the first-order properties of unit magnification and afocality?

To answer this, there is no need to dive into algebraic computation; rather, a single diagram provides the answers.

Consider figure 4.11, where A, A', and O are co-planar, and O is the common centre of the mirrors at B, C, and D, which we call M1, M2, and M3, respectively. We call the line perpendicular to the plane containing O, A and A', which also passes through the points F and P, the axis. F is the principal focus of M1, and P is the principal focus of M2.

If M2 were not present, light from A parallel to the axis would reflect from B through its focus F to A' at an equal-magnitude perpendicular distance from the axis as A (y_p). Similarly, any number of parallel rays to AB would be focussed by M1 at F.

Likewise, in the absence of M2, light travelling from A' parallel to the axis would reflect at D, pass through the long-radius mirror focus at P, and continue to point A. Any number of rays parallel to AD would come to a focus at P. These statements must be true because the plane AOA' contains the centre of curvature of both mirrors, and F and P are the principal foci of the short and long-radius concave mirrors, respectively.

We now define the point of intersection C as the intersection of lines AD and $A'B$. We locate M2, concentric with M1 and M3, such that C is a point on the surface of M2.

We can now see that the triangles ADA$'$ and CAE are similar (opposite angles on parallel lines). Therefore,

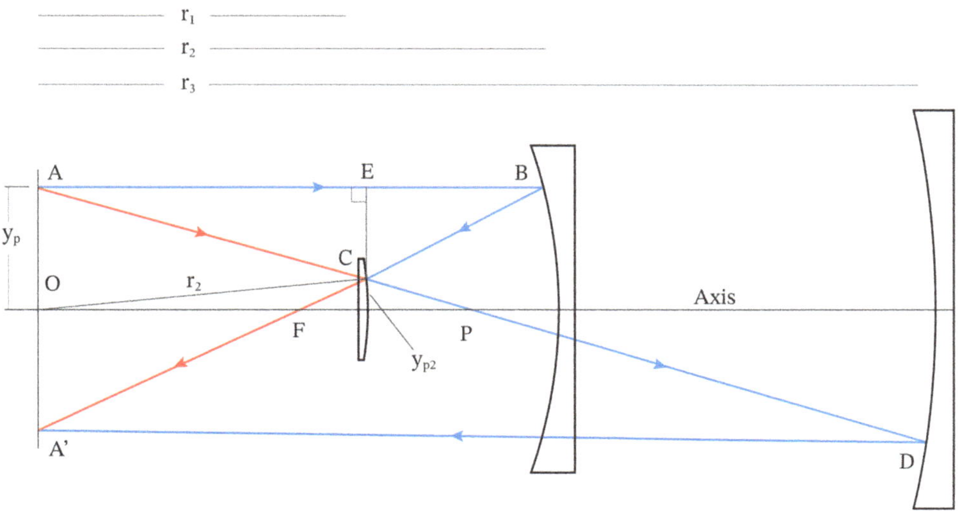

Figure 4.11. In this figure, the blue line represents the path of a ray launched from A parallel to the axis, reflected at *B*, *C*, and *D*, and terminating at the image point *A*'. We can also consider two ray paths in the case where there is no mirror at *C*: A′DA and ABA′.

$$\frac{AA'}{CE} = \frac{A'D}{EA} = \frac{r_3}{r_2}. \tag{4.22}$$

Considering now the triangles ABA′ and EBC, we can see that these are similar (both right-angled and with a common angle). Therefore,

$$\frac{A'A}{CE} = \frac{AB}{EB} = \frac{r_1}{r_1 - r_2}. \tag{4.23}$$

The ratios in equations (4.22) and (4.23) are equal, so we have:

$$\frac{r_3}{r_2} = \frac{r_1}{r_1 - r_2}, \quad \text{or} \quad r_3 = \frac{r_2 r_1}{r_1 - r_2}. \tag{4.24}$$

Therefore, by equation (4.16), this geometry satisfies the Petzval condition. This geometry is general for any spheres M1 and M3 that are concentric. If a secondary mirror M2 is placed such that its surface contains point *C* and it is concentric with M1 and M3, then a similar diagram to figure 4.11 results, and the Petzval sum of the system is again zero.

From this result, we can see that if the system of three mirrors is monocentric but the Petzval condition is NOT met, then the reflected ray DA′ cannot be parallel to the axis, though the ray must come to the same point *A*′.

Therefore, the system in this case would still have unit transverse magnification, but it would no longer be afocal, as incident telecentric light would not be telecentric on exit.

This last fact can easily be seen by considering a pencil of parallel rays parallel to *A*, reflecting off M1 and M2. These rays represent the principal rays from across the

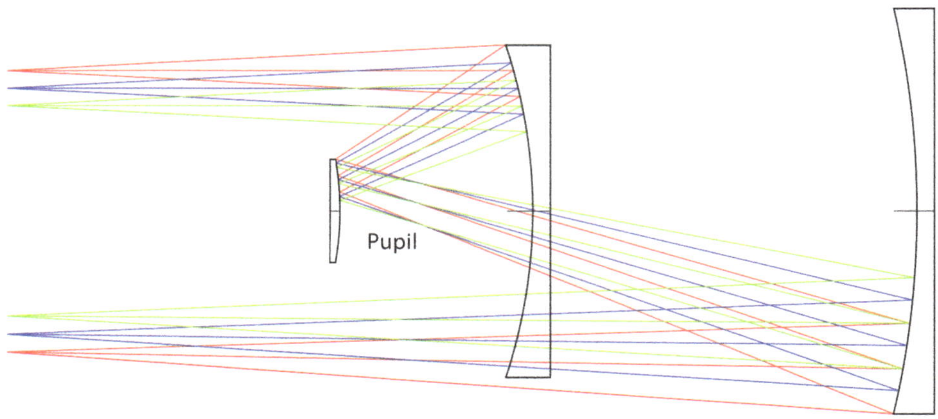

Figure 4.12. The accessible pupil opens up a number of interesting application areas for the Offner relay.

field when the object is telecentric. From the construction, we know that they would come to a focus at point *P*, which is the image of *F* through M2. *P* is also at the focus of M3 in the Petzval-corrected system. If the value of r_3 did not satisfy equation (4.17), M3 would then image P to some finite distance, so the outgoing rays would not be parallel.

Figure 4.12 shows that the pupil of this system, at *P* for a telecentric object, is now removed from the confusion of rays surrounding M2 and is therefore accessible.

When the Offner patent is read carefully, it can be seen that a specific claim is made for exactly the sort of system described here. Offner claims reflecting systems of at least three mirrors, one convex, which are monocentric. Offner points out that such systems are free of spherical aberration and coma when object centred, and further that when the geometric sum of their powers is zero (zero Petzval), the astigmatism is corrected too.

In the large body of literature that describes various incorporations of the Offner relay since its invention, there appears to be very little mention of anything but the classic symmetrical version that is taught in most optics classes. When asymmetry is discussed at all, it is merely to describe the correction of high-order aberrations, as shown in figures 4.7–4.9. In very rare cases, solutions in which the asymmetry is great enough to produce an accessible pupil (free of the confusion of rays around M2) have been found by ray tracing the optical design, and in no case is there any published recognition that such very asymmetric combinations of mirrors have exactly the same optical characteristics as the symmetrical Offner system, to at least the third order.

This lack of general awareness of the asymmetric Offner design in the optical design community is surprising because an accessible pupil is very useful in some applications.

For example, consider an adaptive optics (AO) system. Here, the relay system reimages a telescope focal surface at an instrument. Real-time wavefront error arising from the turbulent atmosphere is measured using light split from within the relay path, and the relay uses this information to provide closed-loop wavefront correction via kilohertz-speed deformable mirrors incorporated in the relay optical path. AO systems commonly use relays based on confocal off-axis paraboloid mirrors, but the non-rotationally symmetrical aspheric surfaces of these are relatively difficult to make and align compared to the spherical surfaces of Offner mirrors. The paraboloidal systems also suffer from residuals of field curvature and distortion.

In 2020, the Gemini North Telescope selected an AO relay based on a modified Offner design supplied by the author for the purposes of a design study [11]. More recently, in a competitive bidding process, a team proposing an Offner-based design won a contract to supply the AO upgrade [12]. The relay design was one of the factors contributing to this outcome.

Another application for such a relay is the Offner spectrograph, first proposed by Mertz [13] and developed by Kwo *et al* [14] and now finding much use in various sensing applications. The traditional symmetrical Offner requires a curved grating to be produced on the secondary mirror, but we can see that a system with an accessible pupil such as in figure 4.12 opens new possibilities for Offner spectrographs incorporating flat transmission or reflection gratings.

A limitation with any Offner spectrograph is that the pupil is quite aberrated, which places certain limitations on performance. Thinking about ways to improve this situation with 'plate-diagram consciousness' led to some very surprising new design insights and a new version of an old idea, the 'perfect instrument' of James C. Maxwell [14], that has also found immediate application. This system is described in the following section.

4.3 A comparison of the Reed and Offner systems

In this section, we will show how an idea arising from 'plate-diagram consciousness' led to a special new class of optical design, which solved an immediate and unanticipated problem in one of the world's largest telescopes.

After completing the work described in the previous section, the author was reflecting on the mistake that had led to the original investigation. That is, the wrong idea that the plates representing the spherical mirrors were at the object surface of the Offner. As we see, the centres of curvature of the spheres lie in this plane, but these centres must be imaged into collimated light space and therefore go to infinity, as we have discussed. Also, in investigating Offner's patent, the author learned for the first time about the Reed pupil relay, as described at the start of this chapter.

The Reed pupil is the Offner object. This optical system has exactly the same geometry of mirrors as the Offner. The only difference is that the object and image conjugates are swapped. The comparison is best summarised in a short table, given here as table 4.1.

As we can see from the table, it is only when we consider the spherical aberration of the pupils of these two systems that the full symmetry becomes apparent. If we consider the Reed in figure 4.1, we see that the rays of the axial pencil for an

Table 4.1. Comparison of Reed and Offner systems: the systems utilise the same components and geometry, being identical. The sole difference is the switch of pupil and object conjugates. For correction, the sole residual aberration of the Offner is not one of the traditional terms but rather an aberration of the pupil. For a telecentric object, the aberration of the parallel pencil of principal rays that occurs with the Offner is identical to the spherical aberration of the Reed.

System	Reed	Offner
Object conjugate	∞ (Fixed)	Plane containing mirror centres.
Pupil	Plane containing mirror centres.	∞ (optional)
M2	Focal surface	Pupil (for telecentric)
Spherical aberration	Yes	No
Coma	No	No
Astigmatism	No	No
Distortion	NO	No
Petzval	No	No
Pupil spherical aberration	No	Yes

infinitely distant object are incident on M1, focus on M2, reflect from there to M3, and are recollimated.

Clearly, the Reed exhibits spherical aberration caused by M1 and M2. In fact, as the 'axial' pencil of the Reed needs to be displaced laterally from the axis to escape self-obscuration, the spherical aberration itself occurs in an off-axis region of the spherically aberrated wavefront. As we have already seen in chapter 2, this means that we have some combination of astigmatism and coma, but as it is coming from a translated pupil on the spherically aberrated wavefront, there is no field dependence to this aberration.

Comparing the Reed with the Offner system in figure 4.4, we see that the marginal rays of the Reed are the principal rays of the Offner when the pupil is at infinity. Therefore, the Offner pupil suffers from exactly the same spherical aberration as the Reed, but this does not affect any of the normal third-order aberrations (though distortion could be discussed as a topic for another book).

The thought occurs, 'What happens if you correct the Reed pupil aberration with a Schmidt plate? Wouldn't that give you both systems? A Reed and an Offner, well corrected simultaneously?'

Yes! Because if you put a Schmidt plate at the object of an Offner system, the marginal rays at the plate have zero height, and so the plate at the Offner object surface is invisible to the Offner. This Schmidt plate WILL correct the spherical aberration of the Offner pupil, which is exactly the Reed spherical aberration.

It was quite an exciting moment to realise this, because somewhere in the dim recesses of memory was a paper by James Clerk Maxwell on what he termed 'perfect instruments' [15] and have also come to be known as 'absolute instruments' [16]. We will take a quick look at that idea now.

4.4 James Clerk Maxwell's perfect instrument

In 1858, James Clerk Maxwell presented a paper on the topic of geometric optics, 'On the general laws of optical instruments' [15]. In it, following a succinct review of the current state of the art, Maxwell decides to consider the formation of optical images from first principles. He first defines the properties of a 'perfect image', paraphrased as follows:

- *Rays departing from each point in the object meet at a single point in the image.*
- *The curvature of the object is exactly matched by the curvature of the image. (Maxwell says that a plane image is produced from a plane object, which is a special-case version of the general curvature condition stated here.)*
- *Geometrical similarity is maintained between the object and the image.*

In the sense of the third-order aberrations described in this book, Maxwell's first condition means that images should be free of spherical aberration, coma, and astigmatism; the second condition means that the instrument introduces no curvature, i.e. the curvature of any object is reproduced in the image; and the third condition means that images should be free of distortion, with no mapping errors.

Using these three conditions, Maxwell next considered whether it would be possible for a single fixed optic, with no moving parts, to simultaneously produce two perfect images of objects at two different distances from the instrument. He was able to show via geometrical arguments that if two perfect images were produced at separated conjugates, perfect images of anything within the field of view of the instrument, i.e. **any** object at **any** conjugate, would be perfectly imaged by such an instrument.

Maxwell then investigated the requirements of such perfect instruments for the general case where the object and image could lie in a space with any refractive index. He proved that, for the case of equal refractive indices in object and image spaces which are both homogeneous optical media (as is the case with the mirrors in air that we consider exclusively in this book), perfect instruments must be 'telescopes' (by which he means 'afocal') and work at unit magnification. The sign of the magnification is unimportant in this context.

Maxwell's proof is given as part of an algebraic argument stretching to several pages, but we show below that some simple geometrical considerations, together with the Lagrange invariant (both in its full form $H = n\, y \sin u$ and in the paraxial limit introduced earlier), suffice to prove these two requirements.

If we consider points on the surface of a hemispherical object, where that surface is concentric with a perfect instrument's entrance pupil, the perfect image of this object must be a hemisphere centred on the exit pupil. The mapping of points on this hemispherical object to its hemispherical image without distortion requires the system to have unit angular magnification, because if this were not the case, then a hemispherical object would map to a hyper-hemisphere or to a 'spherical cap', i.e. a 'perfect image' as defined by Maxwell could not be formed.

The angular invariance of the chief ray between object and image spaces means that a telecentric object, for which the chief rays are parallel to the optical axis, produces a telecentric image, for which the same is true. This means that the system must be 'a telescope', as Maxwell put it, i.e. afocal, because systems that become bi-telecentric when the entrance pupil is at infinity (as must be the case when the chief ray angle is conserved between object and image spaces) are afocal by definition.

Considering further restrictions imposed by the Lagrange invariant in this case, we can immediately see that if a perfect instrument must operate at unit angular magnification, then the transverse magnification must also be unity, from the definition of the invariant.

Maxwell concluded that, apart from a plane mirror, he was not aware of any way in which a perfect instrument could be constructed using conventional optical elements. He went on to show that a perfect instrument could be produced if one were to employ a gradient-index material 'similar to that found in the crystalline lens of a fish' and demonstrated that a 'Maxwell fish-eye' lens with the quadratic gradient he specified was an example of a perfect instrument [17].

4.5 Charles G Wynne and conjugate-shift aberrations

Charles Wynne investigated the third-order aberrations arising from the object-conjugate shift and presented his results in a paper entitled 'Primary Aberrations and Conjugate Change' in 1952 [18]. All factual and quoted information requiring reference in this section is attributable to reference [18]. The results of Wynne's investigations are given below, in Wynne's notation:

$$\begin{aligned}
S_1^* &= S_1 + \bar{\varepsilon}(4S_2 - H\Delta(u^2)) + \bar{\varepsilon}^2(6S_3 - 3H\Delta(u\bar{u})) + \bar{\varepsilon}^3(4S_5 - 3H\Delta(\bar{u}^2)) + \bar{\varepsilon}^4 S_6 \\
S_2^* &= S_2 + \bar{\varepsilon}(3S_3 + S_4 + H\Delta(u\bar{u})) + \bar{\varepsilon}^2(3S_5 + 2H\Delta(\bar{u}^2)) + \bar{\varepsilon}^3 S_6 \\
S_3^* &= S_3 + \bar{\varepsilon}(2S_5 + H\Delta(\bar{u}^2)) + \bar{\varepsilon}^2 S_6 \\
S_4^* &= S_4 \\
S_5^* &= S_5 + \bar{\varepsilon} S_6 \\
S_6^* &= S_6.
\end{aligned} \tag{4.25}$$

In these, the first five S_i are Seidel terms in the usual numerical order (spherical aberration, coma, astigmatism, field curvature, and distortion). The corresponding starred terms for each Seidel aberration are the new values arising from the application of an axial object shift ε.

A new 'Seidel-like' aberration term, S_6, is introduced, which is the spherical aberration of the pupil, as discussed above in the context of the Reed system. This term gives rise to conjugate-shift aberration that solely depends on the magnitude of the shift. Only field curvature has no pupil aberration conjugate-shift dependence, as was also the case with stop shift.

H is the Smith–Helmholtz–Lagrange invariant, u is the marginal ray of the axial pencil angle, $\bar{u}$ is the principal ray angle, and in all cases, the Δ operator refers to the difference between the corresponding angle or angle product values in object space and image space.

Just as we found in the case of stop-shift aberration that we discussed in chapter 2, here we see a hierarchical dependence of Seidel aberrations on object shift. With stop shift, the term with the highest-order angular dependence (distortion) has stop-shift terms that depend on all the lower-order terms. In the case of conjugate shift, this is reversed, so that the terms with the lowest angular dependence have the biggest dependence on other terms.

Wynne made a strong implicit case for the interchangeability of conjugate-shift and stop-shift terms on object/pupil switch, but he never made an explicit formulation of this. A complete theoretical description is tantalisingly close now and could be completely worked out with the right plate-diagram analysis.

With respect to perfect instruments, which have zero conjugate-shift aberration by definition, Wynne showed that his equations could only be completely solved for unit magnification and afocal systems, restating Maxwell's original proof in third-order theory. In such a case, all the delta terms in equation (4.19) become zero, and we are left with the following conditions to solve for zero:

$$\begin{aligned} S_1^* &= S_1 + \bar{\varepsilon}4S_2 + \bar{\varepsilon}^2 6S_3 + \bar{\varepsilon}^3 4S_5 + \bar{\varepsilon}^4 S_6 \\ S_2^* &= S_2 + \bar{\varepsilon}(3S_3 + S_4) + \bar{\varepsilon}^2 3S_5 + \bar{\varepsilon}^3 S_6 \\ S_3^* &= S_3 + \bar{\varepsilon}2S_5 + \bar{\varepsilon}^2 S_6 \\ S_4^* &= S_4 \\ S_5^* &= S_5 + \bar{\varepsilon}S_6 \\ S_6^* &= S_6. \end{aligned} \tag{4.26}$$

Clearly, two conditions are necessary for a system to become a perfect instrument, assuming it is already afocal and has unit magnification. We have already seen in equations (2.40)–(2.42) that stop-shift aberrations are also zero under these conditions.

1) The five normal Seidel aberrations must themselves be zero.
2) The pupil aberration term must be zero.

So, within the boundaries of third-order theory, a perfect instrument has no aberration for any stop position and for any object position. Given the state of the art of optical design at that time, Wynne, echoing Maxwell, concluded that no perfect instrument could be constructed using conventional optical lenses and mirrors and that only the 'trivial solution' exists.

By 'the trivial solution', Wynne most likely meant the flat mirror referred to by Maxwell, as no other qualifying optical system existed as yet in 1952.

4.6 The Schmidt–Offner

Turning our attention back to table 4.1 and referring to equation (4.20), it is now immediately clear that we have the makings of an absolute instrument. All that is required to transform our system into an absolute instrument is to correct the pupil aberration of the Offner and the spherical aberration of the Reed; then, the rhs of equation (4.20) becomes zero.

And, as we concluded in section 4.3, a Schmidt plate concentric with the three mirrors gives exactly the required correction. The Offner correction is unaffected, and the Reed, which was corrected for all aberration except spherical aberration, is now corrected for that as well.

And, as confirmed by both Maxwell and Wynne, we now know that with this system, ANY location of object and stop can be selected without reintroducing third-order aberration—a truly remarkable outcome. In figure 4.13, two large (100 mm) object shifts are introduced into the system depicted in figure 4.12, and the resultant new system is refocussed. Table 4.2 illustrates the aberrations that arise as a result of these object Δ values.

A 200 mm object shift from nominal is introduced (bottom). We see rapidly growing aberration resulting from object-conjugate shift, which is quantified in table 4.2.

Figure 4.13 and table 4.2 show that both Wynne and our own plate-based derivation predict the conjugate-shift aberrations of the relay accurately. Next, in figure 4.14, we consider a system with exactly the same mirrors and positions, but now a Schmidt plate is introduced to correct the pupil spherical aberration.

The correct pupil-correcting Schmidt plate can be calculated in a number of ways. This is the plate that corrects the spherical aberration of the three spherical mirrors of the Reed system. In this case, the marginal ray of the axial pencil must be of a diameter to strike M1 at the same y_1 value as the marginal ray of the most off-axis pencil in our Offner system. A simpler calculation than the one in section 4.2 can be developed because, in this case, we do not have a Δ term to keep track of.

It is also relatively easy to set the system up in ray-tracing code as described in figure 4.14, which is the approach taken here. After the rotationally symmetrical Reed system is produced as shown in figure 4.14, the system is set back to the Offner configuration, now with the plate included (figure 4.15). Table 4.3 updates table 4.2 for the new system with pupil correction included.

Note that the results in table 4.3 were obtained purely by manually changing the object distance in the optical design file and letting the system self-focus with a marginal ray solve. There was no optimisation involved in this result, and nothing moved or changed other than the object and the image. We can conclude that the theory is well validated in practice and that the Schmidt–Offner represents a new class of absolute instrument based on conventional optical components.

On first understanding what this design could do, the author became quite excited. Thorough searches revealed no direct prior art, but the Schmidt–Offner was NOT the first perfect instrument, at least to the third order, constructed from conventional optics. In fact, David Shafer had published papers on this topic and had at least one patent [19–21]. The last Shafer publication on perfect instruments was in 2021 [22]. Of these, one was something akin to a Schmidt–Offner, but it achieved pupil correction with a monocentric meniscus lens concentric with the relay mirrors. Another was a six-mirror system.

The author contacted David Shafer to check whether there was something original here. Shafer commented that he had thought about using a Schmidt plate

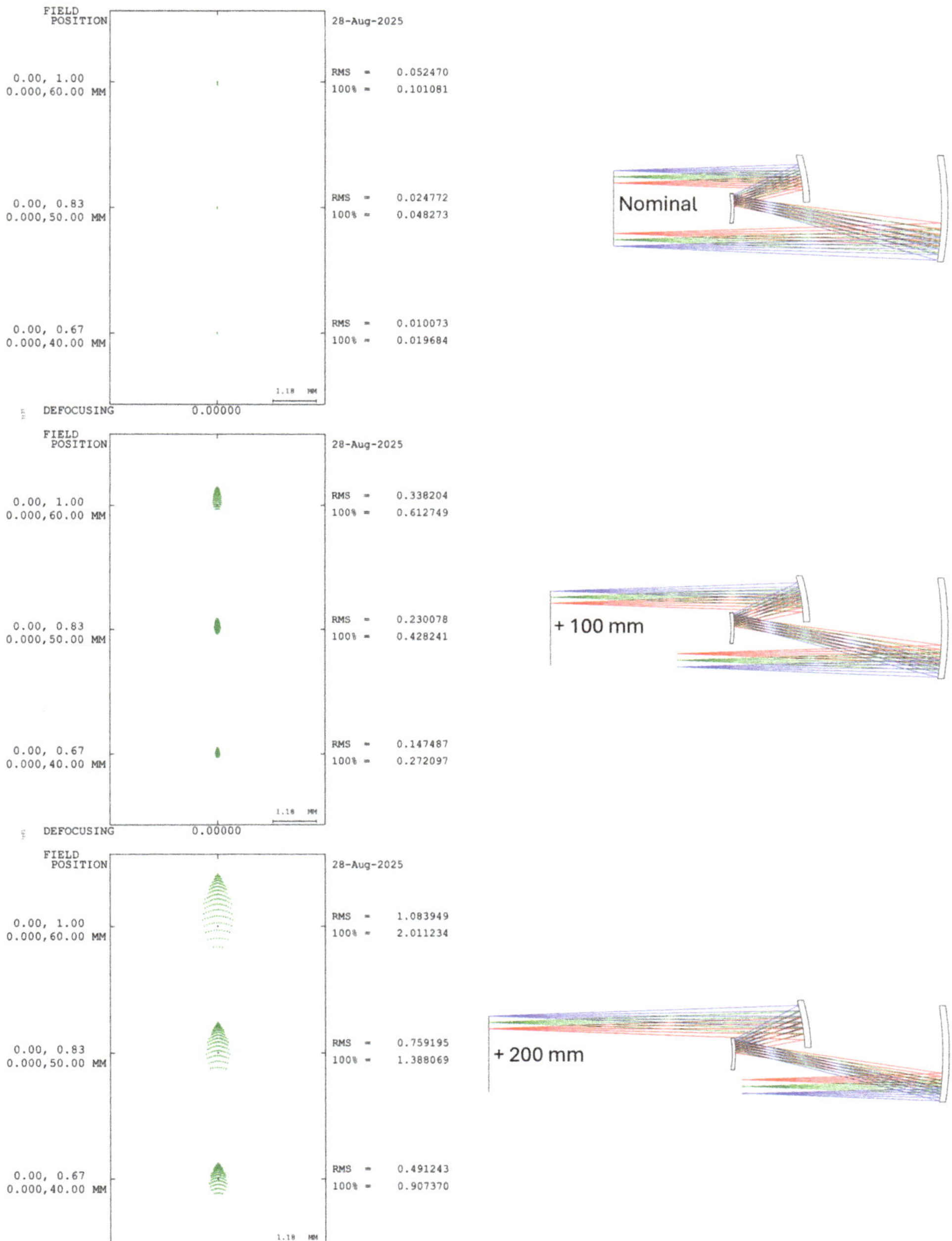

Figure 4.13. Aberrations growing with object-conjugate offset. An unperturbed fully corrected asymmetric Offner system from figure 4.12 (top). A 100 mm object shift is introduced (middle). Created with Code V.

Table 4.2. With several fixed-ratio object-conjugate shifts, we see the power law for conjugate-shift aberration, with coefficients growing as predicted by both equation (4.20) and our earlier investigation (equation (4.15)). Both equations predict that spherical aberration grows as Δ^4, coma grows as Δ^3, and astigmatism grows as Δ^2.

offset	SPHA	COMA	ASTI
0 mm	0.00	0.00	0.00
100 mm	0.01	0.62	9.87
200 mm	0.16	4.94	39.50
400 mm	2.48	39.55	157.98
Ratios			
200:100	16.0	8.0	4.0
400:100	256.0	64.0	16.0

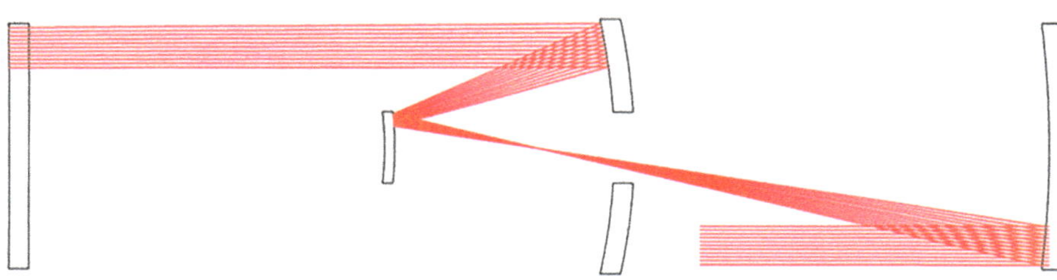

Figure 4.14. To produce the Schmidt plate suitable for our figure 4.12 system, the simplest way is to produce a system with an infinite conjugate that uses a full aperture field stop with a diameter matching that of the largest mirror. The plate strength is easily calculated or solved by ray tracing. To recover our Reed system, we set an off-axis aperture that matches the original Offner field (a 20 mm diameter in this case). Finally, we set the object back at the plane of the mirror centres, recovering an Offner system but now with a pupil-correcting plate. Created with Code V.

but had never got around to practically implementing it and certainly had not written it up [D. R. Shafer, private communication, 12 July 2019].

A patent application was to be prepared and filed later that year (2019) [23]. After the disappointment of realising that Offner's patent actually had fully described asymmetric variants, it was nice to have something to patent after all. It was a fantastic solution, really fantastic, just waiting for the right problem.

And then that problem arrived.

4.7 MAORY/MORFEO and the ELT

The author received a call from the Arcetri Observatory in Florence, Italy, in early 2019, soon after this system had been worked out. The Arcetri team had run into an unexpected problem. Their laser guide stars (LGSs) for the ELT project were suffering from unexpectedly large aberrations in the relay system, and the optical design had become difficult.

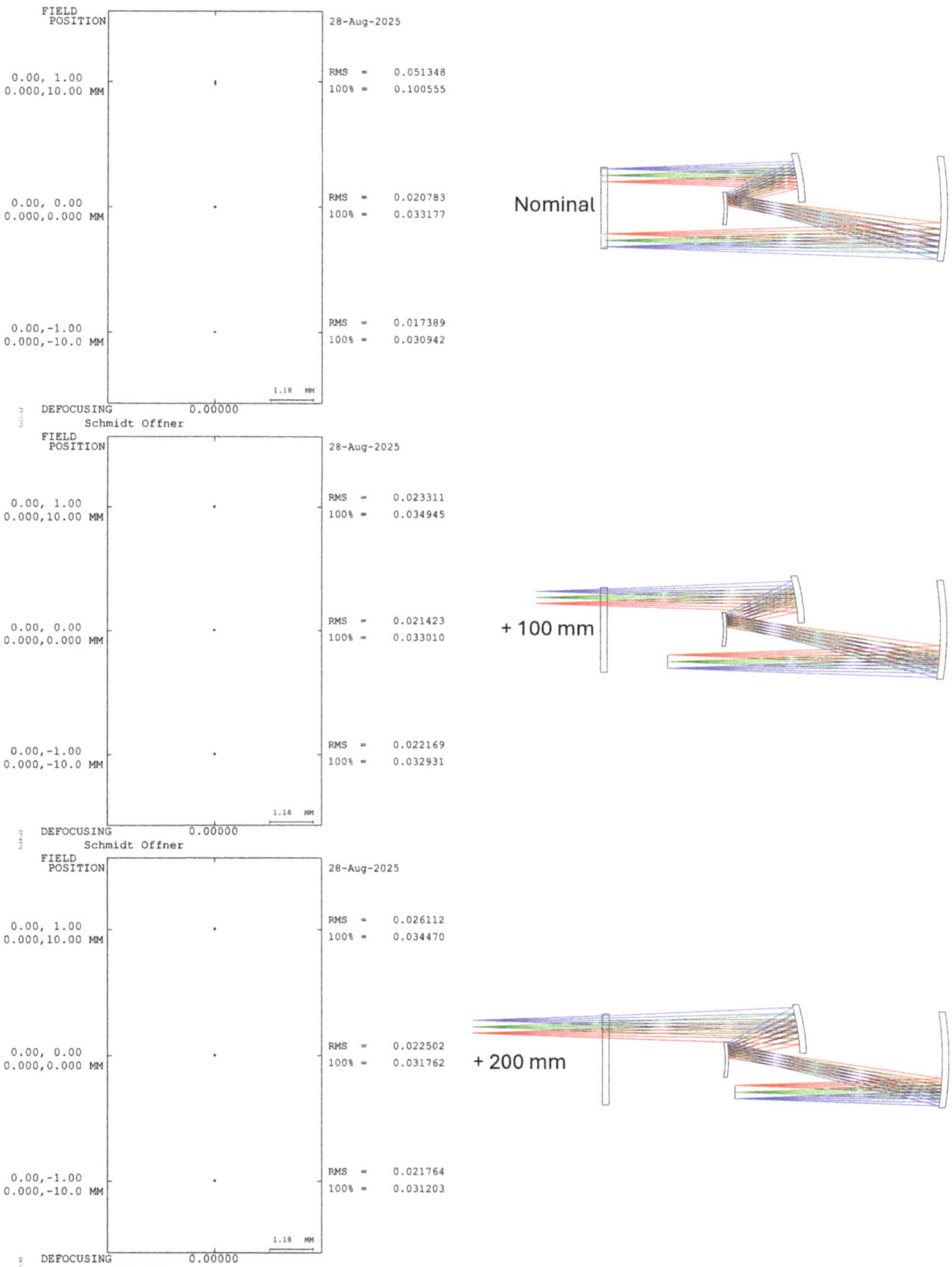

Figure 4.15. With the plate installed and the system reconfigured as an Offner with a finite object, the object-conjugate offsets are repeated. This figure, compared to 4.13, makes it immediately apparent that there is a fundamental change. In this figure, the spot diagrams are deliberately kept to the same scale as in figure 4.13. Created with Code V.

Table 4.3. Updated aberrations with conjugate shift for the Schmidt–Offner. Everything is essentially zero, basically noise with some tiny hints of high-order residual power laws.

Offset	SPHA	COMA	ASTI
0 mm	0.00	0.00	0.00
100 mm	–0.06	–0.02	0.00
200 mm	–0.01	–0.04	–0.01
400 mm	–0.02	–0.08	–0.03
Ratios			
200:100	0.1	2.1	3.3
400:100	0.3	4.5	11.5

The team there, part of the Istituto Nazionale di Astrofisica (INAF), was under contract to the European Southern Observatory to design and construct the AO system, which was at that time called MAORY (Multi-Conjugate Adaptive Optics Relay) [24, 25]. The author was familiar with MAORY from a stint working as technical lead for ELT optics procurement and as an optical designer in the European Southern Observatory (ESO) optics group (2012–15). It should be noted that the name changed. In 2021, out of respect for the cultural rights of the New Zealand Maori people, the name of the relay was changed to MORFEO (Multi-conjugate Adaptive Optics Relay For ELT Observations) [26].

Before going further, it is necessary to describe the optical system in question in more detail. AO systems that incorporate LGSs rely on the simultaneous imaging of objects at two different distances from the telescope: 'science light' in the parlance, meaning light from celestial targets, and 'wavefront sensing light'. When natural starlight is used for wavefront sensing, there is only one object conjugate, which is essentially at infinite distance. However, the 'sky coverage' of AO systems dependent on natural guide stars is severely limited, as the 'guide star' used to read atmospheric turbulence at kHz rates must be bright and must also be close enough to the target that the wavefront of the guide star and that of the angularly separated target are disturbed in essentially the same way by a multi-layered turbulent atmosphere.

To improve on this situation, LGSs have been developed. These are artificial stars produced in the mesosphere, a layer in the upper atmosphere where sodium ions are deposited by burning meteors. A sodium laser of ~20–50 W (continuous wave) is collimated in a small telescope that is boresighted with the observing telescope. The sodium laser light is selectively absorbed and re-emitted by mesospheric sodium atoms, and the re-emitted light radiates in all directions, including directly back down into the telescope aperture. Thus, an artificial star can be produced at any point in the sky, of sufficient brightness to read the atmospheric turbulence at the required rate for real-time atmospheric correction, breaking the limit imposed on *sky coverage* by reliance on natural guide stars.

Whole books can, and have, been written on the subject of laser guide star adaptive optics (LGSAO) [27]. In this book, we raise the subject because of a particular optical problem that arises with LGSAO when telescopes and their associated relays scale up in size from the current state of the art.

The European Southern Observatory's ELT will be the largest in the new generation of extremely large telescopes for the optical and near infrared [28]. The entrance pupil diameter is almost 40 m, which can be compared to the next-largest single-aperture telescopes currently operating: two phased telescopes with 10 m apertures, namely Keck [29] and GTC [30]. This fourfold leap in the primary mirror diameter represents the largest single scaling of the available telescope aperture to have occurred since the earliest days of reflecting telescopes.

It is generally understood that large jumps in scale and complexity can lead to unanticipated problems, particularly in areas in which technological approaches are rapidly evolving. Despite the best attempts of dedicated teams of scientists and engineers, sometimes a problem remains hidden until a later stage of project development suddenly reveals it—the almost-proverbial 'rake in the grass'. The aperture scaling of LGS aberration in a relayed telescope image was one of these problems. If we baseline a 10 m aperture f/16 telescope as being representative of contemporary systems and compare this to the ELT, we can show three types of scaling that we need to consider, which impact the wavefront error of LGSs in the AO relay:

1) Telescope wavefront error for finite conjugates.
2) Absolute image conjugate offset.
3) Conjugate-shift aberrations of scaled relay optics to the same scale as the telescope aperture.

Considering (1) first, if we scale the telescope in aperture, keep the optics spacings to the same scale, as they must remain set for stellar science light, and allow a marginal ray solve to adjust the back focal distance, we can compare the wavefront error we get with a 90 km guide star for the two aperture scales. Table 4.4 gives the results from a model produced in ray-tracing software.

Table 4.4. Aberrations resulting from scaling the telescope aperture for a fixed-distance finite conjugate. Aberrations are given in units of microns.

Aperture vs. Seidel aberrations for 90 km guide star	SPHA	COMA	ASTI
A) 10 m	0.08	0.55	0.16
B) 40 m	4.10	8.75	0.64
Ratios			
B:A	58.6	16.0	4.0

Our expected values for the scaling can be seen by considering our spherical aberration equation (2.28) for a single mirror, repeated here for convenience:

$$W = -\frac{1}{4}nci^2y^2.$$

From this, we expect that

$$c \propto \frac{1}{S}, \; i \propto S, \; \text{and}\, y \propto S \rightarrow W \propto S^3. \tag{4.27}$$

The result in the table from an optical model is close, at 59, given that $S^3 = 64$ here. Recalling also our expressions for coma and astigmatism from equation (2.32), repeated here for convenience,

$$\frac{W}{y_c^4}(4y_p\rho^3\cos^3\theta) \rightarrow -4\frac{W}{y_c^4}ux\left(y_c^3\right) = -4\frac{u}{y_c}x\;W\text{coma},$$

$$\frac{W}{y_c^4}\left(4y_p^2\rho^2\cos^2\theta\right) \rightarrow 4\frac{W}{y_c^4}(ux)^2\left(y_c^2\right) = 4\left(\frac{u}{y_c}\right)^2 x^2W \text{ astigmatism},$$

we can see that the factors of $\frac{1}{y_c}$ explain the reduced power dependence on scale factor for the coma and astigmatism terms and give exactly the power relationships we see in the table:

$$\text{Coma} \propto S^2,$$

$$\text{Asti} \propto S. \tag{4.28}$$

Next, we consider the second identified scaling effect: the shift in image conjugate distance from the principal focus of the telescope. Figure 4.13 gives two formulae which are commonly used to relate object distance and image distance with a thin lens. For our purposes, we will replace our telescopes with thin lenses with the same focal length and calculate the conjugate offsets.

Before we do that, though, in keeping with the spirit of this book in deriving things from first principles, we will quickly develop these first-order formulae, getting in some more practice with our ray-tracing matrix.

1) Thin lens formula.

To start our equation (2.13) matrix, let:

$$(n_0u_0, \; y_0) = (u_0, \; 0),$$

representing an object on the lens axis with an arbitrary marginal ray angle.

The transfer matrix is:

$$A = \begin{pmatrix} 1 & -\frac{t_0}{n_0} \\ 0 & 1 \end{pmatrix} = \begin{pmatrix} 1 & -s \\ 0 & 1 \end{pmatrix},$$

The power matrix we used so far for mirrors changes for a thin lens:

$$B = \begin{pmatrix} 1 & 0 \\ \frac{-2}{r} & 1 \end{pmatrix} \rightarrow \begin{pmatrix} 1 & 0 \\ \frac{1}{f} & 1 \end{pmatrix}, \text{ for a thin lens in air.} \tag{4.29}$$

Note that we replace the power term for a reflecting surface in air derived earlier with the reciprocal focal length, 1/*f*, for a thin lens. Multiplying through gives:

$$\left(n_1 u_1,\ y_0^;\right).\ A.\ B = (u_0,\ 0).\begin{pmatrix} 1 & -s \\ 0 & 1 \end{pmatrix}.\begin{pmatrix} 1 & 1 \\ \frac{1}{f} & 0 \end{pmatrix}$$
$$= u_0\left(1 - \tfrac{s}{f},\ -s\right).$$

Then the image distance from the lens, s', is simply the calculated height divided by the calculated angle, giving

$$s^; = \frac{-s}{1 - \frac{s}{f}} = \frac{fs}{s - f}.$$

Inverting gives the thin lens formula from figure 4.13:

$$\frac{1}{s'} = \frac{s - f}{fs} = \frac{1}{f} - \frac{1}{s} \rightarrow \frac{1}{s} + \frac{1}{s'} = \frac{1}{f}. \tag{4.30}$$

Newton's equation from figure 4.16 is a more useful for our purposes than the thin lens formula. To obtain this from equation (4.24), multiply through by $ss'f$, giving

$$sf + s'f = ss'.$$

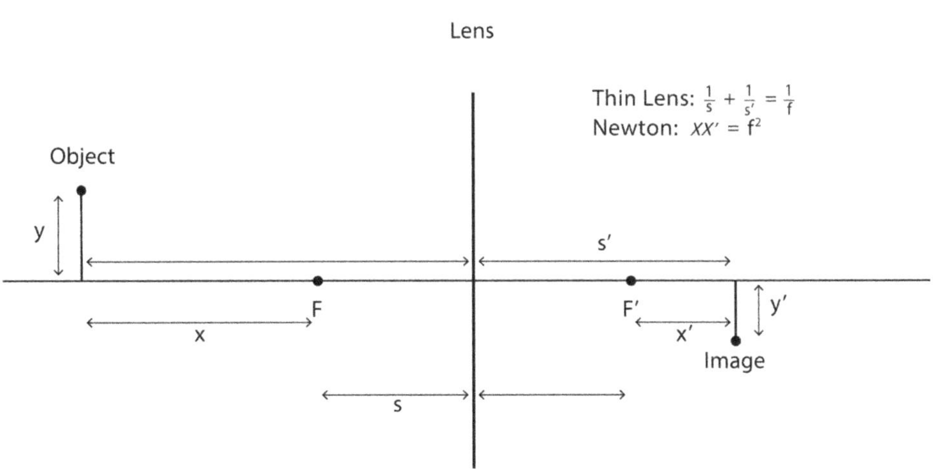

Figure 4.16. Thin lens formula, and Newton's formula, quantities defined.

Rearranging gives:

$$ss' - sf - s'f = 0.$$

Factoring gives:

$$(s - f)(s' - f) = f^2.$$

Substituting the quantities from figure 4.13 gives:

$$xx' = f^2, \tag{4.31}$$

which is the conjugate distance form of the original Gaussian form, or 'Newton's imaging equation'. We use this to populate table 4.5, which presents the image conjugate shifts as a function of telescope scale. The calculations are based on a scale factor of four times relative to the baseline telescope with a 10 m aperture and a focal ratio of f/17.5.

Of course, equation (4.25) already tells us to expect a quadratic relationship between image offset and focal length scale factor, but the table is included to give a feel for the magnitude of shift for LGSs at the ELT scale. We see that 40 m telescopes have 16x the absolute image shift of 10 m telescopes for a given LGS altitude.

We now look at the effects of a scale change on an AO relay. We have seen how an asymmetrical Offner relay can be used in this context, with deformable mirrors located at or around the accessible pupil. An Offner was naturally suited for the European 40 m ELT case because the system required an afocal relay with unit magnification.

We have already calculated the spherical aberration for an object-conjugate shift on an Offner, with the result given in equation (4.12), repeated below:

$$W_{SYS(\Delta)} = -\frac{u_0^4 \Delta^4}{4} \frac{1}{r_1 r_2 r_3},$$

For now, we keep a scale-invariant value for Δ. For the same f-number, u_0 stays constant, and we see that the three radii will scale with a system scale factor S. In this case, the effect of this scaling on system spherical aberration is to reduce it by a factor of

$$W_{SYS} \propto \frac{1}{S^3}. \tag{4.32}$$

Now considering the change in Δ, we expect the absolute value of Δ to grow as S^2. The spherical aberration from equation (4.12) grows as Δ^4, so we therefore expect the scaled system to have spherical aberration increased by

Table 4.5. Image conjugate axial distance offsets from the principal telescope focus, with LGS distance.

	10 m	40 m	Column ratio: 40 m/10 m
Focal length for f/17.5 (m)	175	700	4
Focal surface offset for a 90 km guide star	340.9	5487	16
Focal surface offset for a 150 km guide star	205.1	3281	16

$$(S^2)^4 = S^8. \tag{4.33}$$

These effects multiply, so the final expected change in spherical aberration for a scaled Offner relay follows

$$W_{SYS} \propto S^5. \tag{4.34}$$

We can check this against the model that was used to produce table 4.2. Table 4.6 begins with data from the model set with a 50 mm object axial position offset; it then applies a 4x scale to the optical system and then a 16x scale to the original object offset.

Considering the A: B ratio first, we see the expected $\frac{1}{S^3}$ scaling when the Offner relay size is increased with a fixed absolute focus offset and fixed input numerical aperture.

Next, for the C: B ratio, we see the expected S^8 scaling due to focus offset. The absolute value of the object-conjugate shift itself scales as S^2, then the aberration grows as S^2 taken to the 4th power.

Finally, with the C: A ratio, we see the resultant S^5 scale for spherical aberration, reducing to S^4 for coma and S^3 for astigmatism.

This result combines the scaling effects (2) (focus offset scaling) and (3) (relay scaling). It remains to multiply this result with (1) (feed-telescope scaling).

Doing this yields, for the scaling of aberrations in a telescope combined with an Offner relay imaging a fixed finite-distance conjugate:

$$\begin{aligned} \text{spherical aberration} &\propto S^8, \\ \text{coma} &\propto S^6, \\ \text{astigmatism} &\propto S^4. \end{aligned} \tag{4.35}$$

Table 4.6. The A:B ratio verifies the reduction of aberrations predicted by equation (4.26). The ratio C:B verifies the scaling of aberrations with an S^2 conjugate shift only, where $S = 4$ here. The ratio C:A gives the combined effect of scaling the relay optical system by S and the object axial offset by S^2. Aberrations are in units of microns. Note that while astigmatism has the weakest power growth with scale, it is the dominant residual to start with and so the 'worst offender'.

Offset	SPHA	COMA	ASTI
A) 50 mm offset base	0.000 604	0.067 58	1.889 92
B) 50 mm offset, scaled up by 4x	9.4399E-06	0.003 7713	0.3766
C) 800 mm offset, scaled up by 4x	0.62	15.45	96.42
Ratios			
A:B	64.0	17.9	5.0
C:B	65 536.7	4096.0	256.0
C:A	1024.3	228.6	51.0

In practice, while the spherical aberration scaling term has the highest power dependence, the astigmatism residual is by far the dominant starting term for off-axis relays (Row A, table 4.6). At the level of scaling considered here, astigmatism remains the dominant wavefront error associated with conjugate shift, as was shown in Row C, table 4.6.

While the implications of (1) LGS aberration arising from scaling the telescope for aperture and (2) the quadratic dependence of focal shift with scale were well understood in advance and, on their own, caused manageable wavefront error, (3) the scaling of conjugate-shift aberration in the scaled relay is quite a subtle effect that would have been difficult to spot at all in the 1–10 m aperture scales of current adaptive telescopes. This effect really only became apparent when the INAF team began the detailed optical design process of the relay.

The large finite-conjugate aberration is not a fixed aberration, as the AO system has to work with a range of guide star altitudes from 90 to 180 km, so the amount of aberration is expected to vary constantly during observation. This would normally have necessitated some form of dynamic correction in the optical path of the LGS.

Interestingly, the perfect solution—a Schmidt–Offner—had just been found and was in the process of being patented just as the magnitude of this problem became apparent. The author took on a small design support contract with INAF in 2019 and produced what became the baseline design for MAORY, as described in references [24–26].

The patent application, when it was submitted, used the broader terms recommended in patenting. It claimed relevance to:

- any optical relay system that was *substantially* afocal and of *substantially* unit magnification, and
- was *substantially fully corrected* to at least the third order of aberration, and
- had its pupil aberration corrected by means of a refracting or reflecting aspheric plate in the near region to the focus.

These claims recognised that there are any number of combinations of optics that can produce unit magnification, afocal flat-field, distortion-free anastigmats, which can then be turned into perfect instruments via the aspheric plate near focus.

In fact, by the time MAORY had become MORFEO, a new base relay design inspired by Kutter's Schiefspiegler [31], popular in the amateur astronomy community, had been provided by ESO and replaced the original Offner system, predominantly because it could be made more compact. The new design still employed the plate and was another embodiment of the general class of perfect instrument systems described in the patent application [23]. The final design of MORFEO is described in reference [32].

This chapter has presented technical information and has also told a story. It is hoped that the technical information is found useful by readers and that the story helps, in part, to answer the question that is sometimes asked by optical designers who know about the plate diagram: 'Why should I take the trouble to learn this?'

A significant answer is that plate-diagram-inspired curiosity eventually revealed the broader class of Offner configurations that had gone unrealised for decades.

The discovery of the remarkable properties of the Schmidt-corrected relay is another answer. The fact that these systems found immediate application in competitively bid multi-million-dollar projects, proving that they were, in fact, superior solutions to those previously available, is a third. Overall, it should be clear that the author's claim, echoing Burch, that the plate diagram offers designers fresh new ways of looking at things, ways that can enable significant new discoveries, is vindicated by these results.

In the final chapter, we will discuss how analytical methods combined with modern computing power open up entirely new ways of finding optical solutions; these new methods have barely begun to reach their potential.

References

[1] Offner A 1973 *US Patent* No. 3,748,015, filed September 8, 1971, issued July 24

[2] Reed W A 1965 *US Patent* No. 3,190,171, filed February 26, 1962, issued June 22

[3] Rakich A 2017 Reflecting anastigmatic optical systems: a retrospective *Int. Optical Design Conf. 2017 (Optical Design and Fabrication 2017 (Freeform, IODC, OFT)), Paper ITh1B.4, OSA Technical Digest (online)* (Optica Publishing Group)

[4] Brandl B *et al* 2024 Final design and status of METIS, the mid-infrared ELT imager and spectrograph *Proc. SPIE 13096, Ground-based and Airborne Instrumentation for Astronomy X* 1309612

[5] Thatte N *et al* 2024 HARMONI at ELT: project status and instrument overview *Proc. SPIE 13096, Ground-based and Airborne Instrumentation for Astronomy X* 1309610

[6] Agócs T *et al* 2016 Preliminary optical design for the common fore optics of METIS *Proc. SPIE* **9908** 99089Q

[7] Rakich A 2020 Email to T Agócs, subject: 'Offner relay paper/broken-symmetry Offner suggestions for METIS and HARMONI,' 6 March 2020 (personal communication)

[8] Agócs T 2020 Email to A Rakich, subject: 'Re: Offner relay / broken-symmetry Offner in METIS CFO,' 10 March 2020 (personal communication)

[9] Rakich A and Rogers J R 2020 A generalized Offner relay with an accessible pupil *Proc. SPIE 11451 Advances in Optical and Mechanical Technologies for Telescopes and Instrumentation IV 114510B*

[10] Rogers J R and Rakich A 2020 The importance of Petzval correction in generalized Offner designs *Proc. SPIE 11482 Current Developments in Lens Design and Optical Engineering XXI 1148207*

[11] Chirre E, Sivo G, Scharwächter J, Andersen M, Provost N, Marin E, van Dam M, Chinn B, Cavedoni C *et al* 2020 Potential optical design of a multi-conjugate adaptive optics instrument based on a modified Offner concentric relay for Gemini North *Proc. of SPIE 11448: Adaptive Optics Systems VII* (SPIE) 114487F

[12] Jouve P *et al* 2024 AOB: the new adaptive optics bench at Gemini North *Proc. of SPIE 13097: Adaptive Optics Systems IX* (SPIE) 130976C

[13] Mertz L 1977 Concentric spectrographs *Appl. Opt.* **16** 3122–4

[14] Kwo D, Lawrence G and Chrisp M P 1987 Design of a grating spectrometer from a 1:1 Offner mirror system *Proc. SPIE* **818** 275–81

[15] Maxwell J C 1858 On the General Laws of Optical Instruments *Q. J. Pure Appl. Math* **2** 233–46

[16] Miñano J C 2006 Perfect imaging in a homogeneous three-dimensional region *Opt. Express* **14** 9627–35

[17] Beech M 2011 Hooke's dream lens, Maxwell's Fish-eye, and Luneburg's sphere *Bull. Sci. Instrum. Soc.* **109** 2–8

[18] Wynne C G 1952 Primary aberrations and conjugate change *Proc. Phys. Soc. Lond. Sect.* B **65** 429–37

[19] Shafer D R 1991 New perfect optical instument *OSA Annual Meeting 1991 (San Jose, CA, 3–8 Nov. 1991)* (Optical Society of America)

[20] Shafer D R 2005 Some odd and interesting monocentric designs *Proc. of SPIE 5865: Tribute to Warren J. Smith: A Legacy in Lens Design and Optical Engineering* (SPIE) 586508

[21] Shafer D R 1987 Ring field projection system *US Patent* 4,711,535

[22] Shafer D R 2021 A perfect lens design hiding in plain sight for 167 years *Int. Optical Design Conf. 2021 Proc. SPIE* **12078** 12078P

[23] Rakich A 2020 Optical systems capable of forming highly corrected images of 3-dimensional objects *International Patent Application* No. PCT/NZ2020/050105

[24] Magrin D *et al* 2020 MAORY: optical configuration and expected optical performances *Proc. SPIE 11448, Adaptive Optics Systems VII* 1144834

[25] Ciliegi P *et al* 2021 MAORY: a multi-conjugate adaptive optics relay for E-ELT *ESO Messenger* **182** 13–6

[26] Busoni L *et al* 2023 MORFEO enters final design phase, adaptive optics for extremely large telescopes 7 (AO4ELT7), Avignon, France HAL: hal-04429120

[27] Tyson R K and Frazier B W 2022 *Principles of Adaptive Optics* 5th edn (Boca Raton, FL: CRC Press)

[28] Gilmozzi R and Spyromilio J 2007 The European extremely large telescope (E-ELT) *The Messenger* **127** 11–9 https://messenger.eso.org/127/

[29] Nelson J E and Gillingham P R 1994 Overview of the performance of the W. M. Keck observatory *Proc. SPIE* **2199** 82–93

[30] Álvarez P *et al* 2010 The GTC project: from commissioning to regular science operation. Current performance and first science results *Proc. SPIE* **7733** 773305

[31] Kutter A 1951 Spiegelteleskop mit schief gestellten optischen Achsen *German Patent* DE845635C (filed 6 Nov 1951, published 28 Aug 1952)

[32] Busoni L *et al* 2023 MORFEO enters final design phase, adaptive optics for extremely large telescopes 7 (AO4ELT7), Avignon, France. HAL: hal-04429120

IOP Publishing

Analytical Lens Design using the Optical Plate Diagram
An introduction to the fundamentals with practical applications
Andrew Rakich

Chapter 5

System surveys

At the beginning of this book, we introduced the concept, first given by Aldis [1], that to achieve an anastigmatic optical system, a minimum of four non-coincident plates would be required (see chapter 1, p 1-7). In chapter 3 (pp 3-6–3-10), we provided a version of Burch's proof of this 'see-saw' correction.

It is interesting to consider this with respect to the available degrees of freedom in optical design. In the case of reflecting telescopes, we can look at the simplest possible systems that can produce four non-coincident plates; there are three of them:

- two aspheric mirrors,
- three mirrors (one aspheric), and
- four spherical mirrors.

In this chapter, we will look at each of these in turn and see how a combination of plate-diagram thinking, analytical solutions, and modern computing power suggests alternative approaches to optical design that can complement and enhance results obtained purely by ray tracing and optimisation.

5.1 Two-aspheric-mirror anastigmats

There is a rich history of research and development in the field of two-mirror anastigmats. This book has already discussed various contributions, in particular, those of Schwarzschild [2], Burch [3], and Couder [4]. There were significant developments in the theory in the 1950s and 1960s, when even the great Paul Erdös authored a paper on concentric spheres [5]. The 'definitive triumph' was probably Wynne's 1969 paper, which, with the benefit of all previous work, presented a systematic definition of every possible solution [6].

We have already looked at some of these solutions, considering both a single anastigmat of the concave–concave mirror type, discovered by Schwarzschild and

doi:10.1088/978-0-7503-3099-2ch5

named for Couder, and also the two-spherical-mirror anastigmat with a convex primary mirror and a remarkable geometrical relationship to the golden section.

In this section, we present a simple survey of all possible solutions with one infinite conjugate to demonstrate both how readily this is achieved and how useful such an approach is for drawing quick conclusions about a whole class of systems and identifying interesting special cases.

It is possible to formulate expressions in terms of the primary mirror radius r_1, and the mirror separation t_1, which can be solved for r_2, k_1, and k_2, where the latter two quantities are the conic constants for their indexed mirrors. For convenience, we choose an aperture stop on the surface of M1, so all four x_i are measured from that, and x_{a1}, representing the asphericity of M1, equals zero.

We now consider a ray trace for light entering the system.

Let refractive indices begin with $n_1 = 1$ and alternate sign on reflection:

$$\text{marginal ray of axial pencil: } (0,\ y_1)$$

$$\text{power matrix for M1: } A = \begin{pmatrix} 1 & 0 \\ \dfrac{-2}{r_1} & 1 \end{pmatrix}.$$

$$\text{first transfer matrix: } B = \begin{pmatrix} 1 & -\dfrac{t_1}{n_2} \\ 0 & 1 \end{pmatrix} = \begin{pmatrix} 1 & t_1 \\ 0 & 1 \end{pmatrix}.$$

$$\text{power matrix for M2: } C = \begin{pmatrix} 1 & 0 \\ \dfrac{-2}{r_1} & 1 \end{pmatrix}.$$

$$(n_2 u_2,\ y_2) = (0,\ y_1).\ A.\ B = \left(\frac{2y_1}{r_1},\ y_1\left(1 - \frac{2t_1}{r_1}\right)\right).$$

$$\varphi_2 = \frac{y_2}{r_2} = \frac{y_1(r_1 - 2t_1)}{r_1 r_2}.$$

$$i_2 = \varphi_2 - u_2 = \frac{y_1(r_1 - 2(r_2 + t_1))}{r_1 r_2}. \tag{5.1}$$

5.2 Ray trace for M2 surface vertex imaged through M1 (x_{a2})

(Note: we first trace a ray from the M2 surface vertex from left to right towards M1, so we make the refractive index $n_1 = 1$ again in this case. As t_1 was previously defined as the distance from M1 to M2, it follows that going from M2 to M1, that distance is $-t_1$).

$$\text{ray on M2 surface vertex (on } - \text{ axis) with arbitrary angle: } (n_1 u_1,\ 0)$$

$$\text{first transfer matrix } \alpha\text{:}\quad \alpha = \begin{pmatrix} 1 & -\dfrac{-t_1}{n_1} \\ 0 & 1 \end{pmatrix} = \begin{pmatrix} 1 & t_1 \\ 0 & 1 \end{pmatrix}.$$

$$\text{power matrix for M1, as before:}\quad A = \begin{pmatrix} 1 & 0 \\ \dfrac{-2}{r_1} & 1 \end{pmatrix}.$$

Tracing the ray from the M2 vertex through reflection in M1 gives

$$(n_2u', y_+) = \alpha.\, A = \left(n_1u_1\left(1 - \frac{2t_1}{r_1}\right),\ n_1u_1.\, t_1\right),$$

from which we obtain

$$x_{2a} = \frac{y_+}{u'} = \frac{r_1t_1}{2t_1 - r_1}. \tag{5.2}$$

5.3 Ray trace for M2 centre of curvature imaged through M1 (x_{s2})

We can take a shortcut here, noting that the centre of curvature is at a distance r_2 from the M2 surface vertex, so we can simply substitute $(t_1 + r_2)$ for t_1 in equations (5.2), giving

$$x_{s2} = \frac{r_1(t_1 + r_2)}{2(t_1 + r_2) - r_1}. \tag{5.3}$$

Using these quantities, we can calculate the spherical aberration for each plate, letting k_i denote the mirror conic constants and using quantities derived in equations (5.1):

$$W_{1s} = -\frac{1}{4}n_1\frac{1}{r_1^3}y_1^4 = -\frac{y_1^4}{4r_1^3}.$$

$$W_{1a} = k_1 W_{1s} = -k_1\frac{1}{4}n_1\frac{1}{r_1^3}y_1^4 = -\frac{k_1y_1^4}{4r_1^3}.$$

$$W_{2s} = -\frac{1}{4}n_2\frac{1}{r_2}i_2^2y_2^2 = \frac{(r_1 - 2t_1)^2(r_1 - 2(r_2 + t_1))^2y_1^4}{4r_1^4r_2^3}.$$

$$W_{2a} = -k_2\frac{1}{4}n_3\frac{y_2^4}{r_2^3} = \frac{k_2\left(y_1 - \frac{2t_1y_1}{r_1}\right)^4}{4r_2^3}. \tag{5.4}$$

With the results of equations (5.2), (5.3), and (5.4), and noting that $x_{1s} = r_1$ and $x_{1a} = 0$due to our choice to place the stop on M1, we can formulate our plate equations:

$$\sum_1^2 W_{is}+\sum_1^2 W_{ia} = \frac{(-r_1r_2^3 - k_1r_1r_2^3 + k_2(r_1 - 2t_1)^4 + (r_1 - 2t_1)^2(r_1 - 2(r_2 + t_1))^2)y_1^4}{4r_1^4r_2^3} = 0$$

$$\sum_1^2 x_{is} W_{is}+\sum_1^2 x_{ia} W_{ia} = y_1^4 F_1(r_1, r_2, k_1, k_2, t_1) = 0,$$

$$\sum_1^2 x_{is}^2 W_{is}+\sum_1^2 x_{ia}^2 W_{ia} = y_1^4 F_2(r_1, r_2, k_1, k_2, t_1) = 0. \tag{5.5}$$

F_i indicate functions of the bracketed quantities that are too long to write out here but which can be found in appendix 2 in the Mathematica script for this section. We note that y_1^4 is common to all three equations and so can be immediately dropped. If we solve these equations simultaneously for r_2, k_1, and k_2, we obtain the following expressions representing two independent solutions, in terms of t_1 and r_1:

$$\{r_2 \to r_1 - 2t_1, k_1 \to -1, k_2 \to -1\},$$

and

$$\left\{r_2 \to \frac{-r_1t_1 + 2t_1^2}{r_1 - t_1}, k_1 \to \frac{-r_1^3 + 2r_1^2t_1 - t_1^3}{t_1^3}, k_2 \to \frac{-r_1^2t_1 + r_1t_1^2 + t_1^3}{(r_1 - t_1)^3}\right\}. \tag{5.6}$$

The first solution, with $r_2 = r_1 - 2t_1$ and conic constants of -1, represents the Mersenne afocal anastigmats with which we are already familiar. In the case of the second solution, we can assign an arbitrary value to r_1 and evaluate the given functions over some range of t_1. We do not lose any generality in doing this, as the resultant anastigmatic systems can be scaled to any value of r_1.

Figures 5.1 and 5.2 present these results for two cases. In figure 5.1, the primary mirror is concave. For every value of t_1, the corresponding values of r_2, k_1, and k_2 produce a two-mirror anastigmat. In both figures, negative values of t_1 represent systems with real air space, and positive values represent systems with a virtual M2.

We can immediately see some interesting features in this figure. The radius r_1 is -1000 mm, so the focus for M1 is at -500 mm. M2 has a positive radius over that range, so the mirror is concave to M1, and curvature is large and grows asymptotically as the mirror spacing approaches zero. The Couder solutions lie in this region, with concave secondary mirrors before the prime focus. We can see a solution at around -600 mm, where the two conic constant curves simultaneously become zero. In fact, the value of t_1 here is $\phi - 1$, and again we have monocentric spheres with the golden ratio throughout the design geometry. Beyond this, there is a solution where the mirror curvature goes to zero. Finally, beyond that, there is a solution where the conic constant of M1 again goes to zero.

With a solution set defined like this, it is a simple matter to filter the set for interesting or viable solutions. For example, in figure 5.2, we plot some basic data for an optical system, having first selected a value for y_1 of 125 mm.

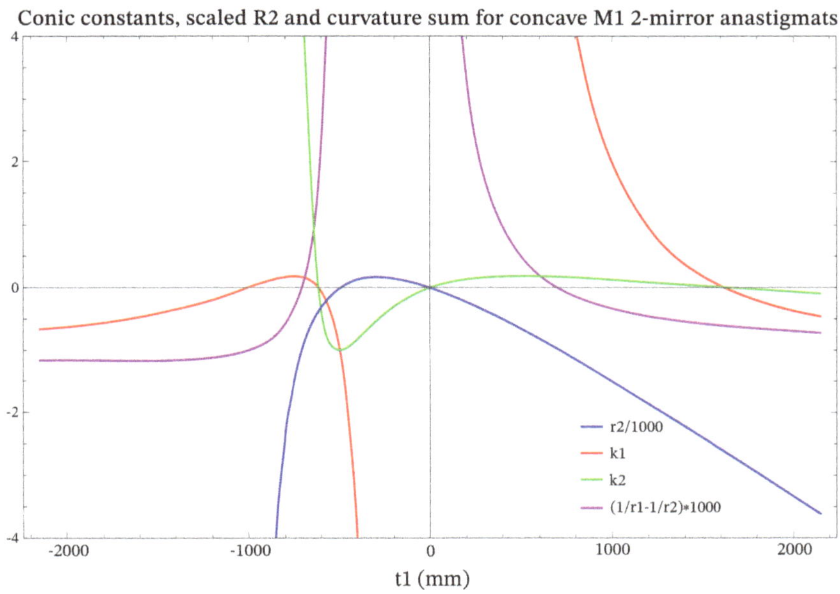

Figure 5.1. Solutions with negative values of t_1 have a real air space from M1 to M2, and those with positive values have a virtual air space. Virtual air space solutions are discussed below.

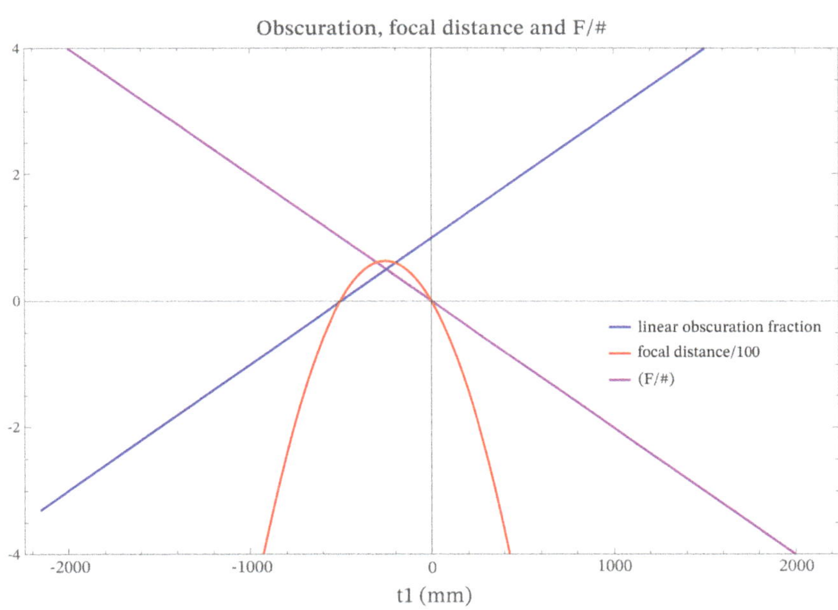

Figure 5.2. Various parameters of interest can be evaluated graphically or algorithmically.

We can see from figure 5.2 that the only solutions that have a real air space to M2 and a real image lie in the region $-500 < t_1 < 0$. We zoom in on that region in figure 5.3.

Selecting the $t_1 = -400$ mm solution, we get:

$$r_2 = 133.33 \text{ mm}$$

$$k_1 = -4.125$$

$$k_2 = -0.8148$$

$$y_2 = 25 \text{ mm}$$

$F/\# = 0.8.$

This solution, an anastigmat, was checked using ray-tracing software. With all calculated parameters correct, it is shown in figure 5.4.

Solutions with a convex M2 lying beyond the prime focus have virtual images, but these can still be useful systems. For example, David Shafer showed how such a system with a virtual anastigmatic focus can be added to another anastigmatic system that conjugates to that focal position [7]. In reference [7], Shafer used the virtual image 'golden' system of monocentric spheres shown in figure 5.5, together with a concentric spherical-mirror pair with two finite conjugates, to produce a four-spherical-mirror unobstructed anastigmat.

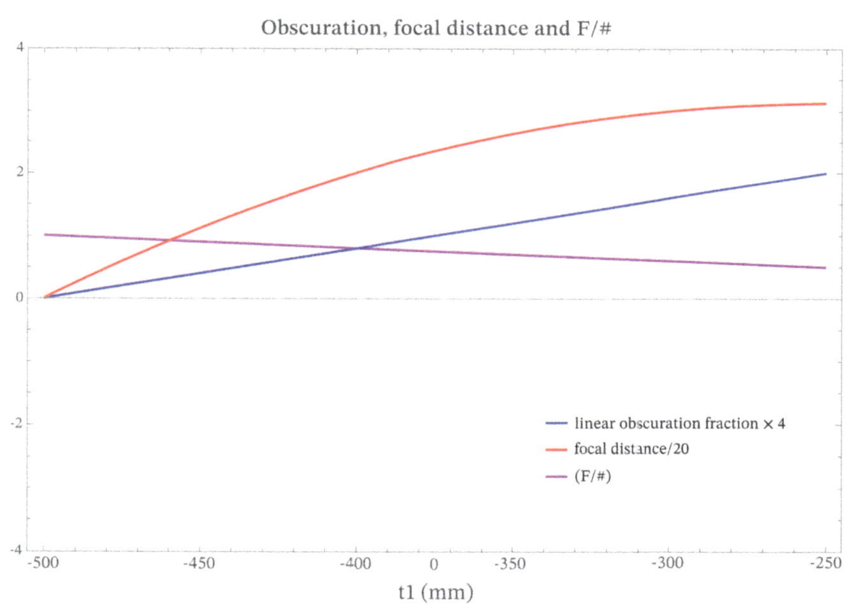

Figure 5.3. Zooming in on a region of practical solutions, we could potentially map any number of characteristics, such as degree of asphericity, maximum incidence angles (high incidence angles can be indicative of a poorly conditioned system with high-order aberrations), etc. Here, we can see that at $t_1 = -400$ mm, we have a solution with ~20% linear obscuration at ~F/0.8.

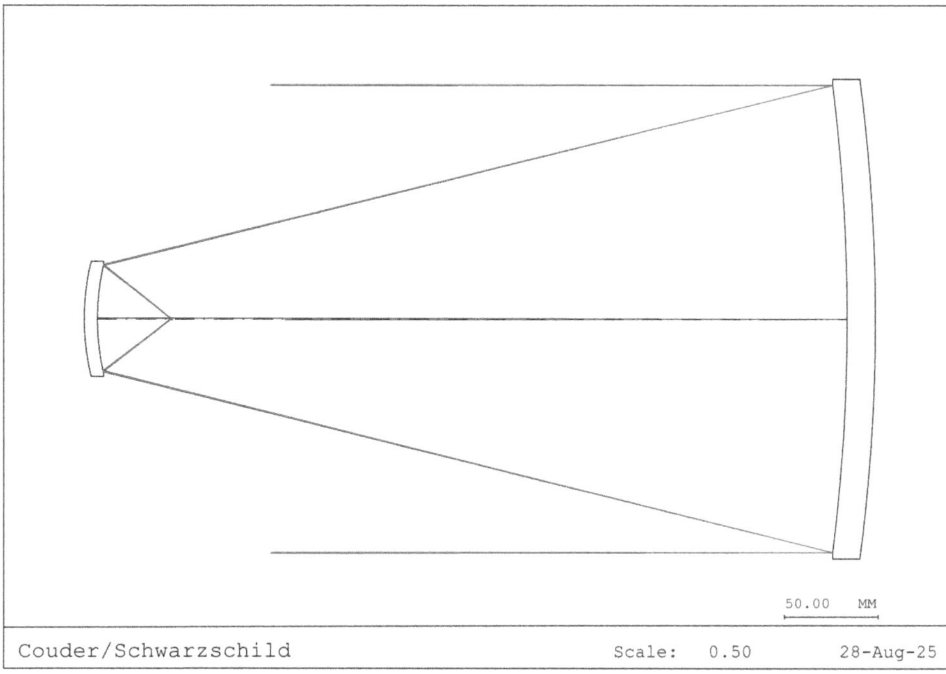

Figure 5.4. Concave anastigmat of the Couder/Schwarzschild type with a large field curvature. Created with Code V.

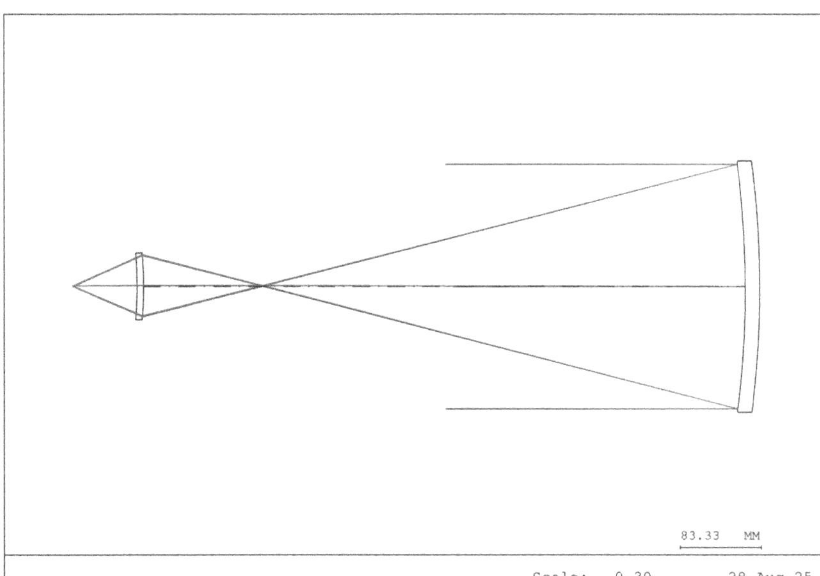

Figure 5.5. Virtual image, all-spherical-mirror anastigmat. Created with Code V.

Given that such systems with virtual foci are potentially useful, it is worth considering the final interesting point noted above, i.e. the flat-field solution with a virtual image. On its own, it is not of much use, but as a potential building block of 'anastigmat algebra', it is worth bearing in mind. Before we turn to the convex M1 solutions, it is worth considering the solutions in which t_1 is positive, representing a virtual air space. As with the virtual-image solutions just discussed, while these systems on their own are not useful, they could possibly be useful in combination with other systems, and the flat-field and spherical-mirror solutions there are worth keeping in mind for the potential future development of this idea of building anastigmats from simple anastigmats.

Another thing to consider is that in the discussion of the Paul system, we saw that a number of different real-world optical systems can have identical plate systems. Another unexplored area in this analytical space is how useful it is to start from an impractical optical system, convert it to a plate system, and then explore all the combinatorial possibilities of real optical systems that produce the same plate diagram. This would yield a tractable, finite number of possibilities, and automated routines could very quickly sift through all the possibilities, isolating the interesting ones (something like panning for gold).

Turning to convex primary solutions, we see the solution curves given in figure 5.6.

The spherical-mirror solution was evaluated in chapter 3 (the 'golden' solution). The flat-field two-mirror anastigmat is described in detail by Wynne [6], who gives a

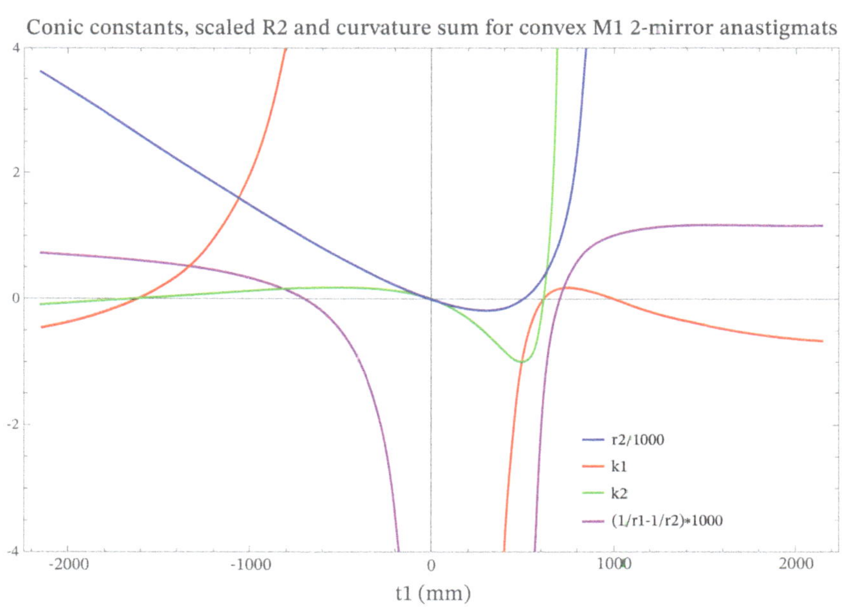

Figure 5.6. As with the concave primary solutions, the only 'real' solutions have negative values of t_1. The point where the curvature goes to zero (flat-field $t_1 \approx -600$) and the point where the mirrors become spherical ($t_1 \approx -1.681$) are of interest.

version corrected analytically to the 5th order with a high numerical aperture, wide field, and excellent performance. A more recent example of a similar approach can be found in Sasian [8].

Before leaving this section, there are several points to consider. Unlike a ray-tracing solution, this relatively simple mathematics has defined an entire class of systems. All possibilities are uniquely identified. To achieve this result with ray tracing would be an exhaustive slog of varying parameters and re-optimising. While such approaches can be and have been undertaken, for example, in various global optimisation strategies, these brute-force ray-tracing methods are orders of magnitude less computationally efficient than analytical solutions and still do not completely define a solution space in the same way as can be done analytically.

A final point to bear in mind vastly extends the scope of two-mirror systems. In this section, we considered only systems with one infinite conjugate, telescopes. There are an infinite number of similar solutions for systems with two finite conjugates. This is discussed by Wynne [6] and Erdös [5] and is the subject of a patent by Kavanagh [9].

A broad general analytical survey using modern computing power would be relatively straightforward to set up using the methods described above, with the addition of a 'paraxial lens' to produce a collimated light space for the plates in object space. That is, for any finite-conjugate system, plates can be calculated to a single optical space anywhere in the system, then reimaged by a transform that collimates the light in that space. This results in a correctly spaced plate diagram in any convergent or divergent light space.

While there is still a lot to be solved analytically in two-mirror systems, especially with respect to the concept of producing an 'anastigmat building block set' of simple systems, the overall geometry is quite limited with just two mirrors. Even so, from 1905 to modern times, papers have continued to be written by people exploring or using this design space. A much more limited body of analytical work exists for the case of three-mirror anastigmats. The next section presents an approach to surveying these.

5.4 Survey of simplest possible three-mirror anastigmats

This section introduces an approach that can be thought of as extending the work of Schwarzschild and Wynne to three-mirror anastigmats. As was the case with two-mirror anastigmats, analytical expressions can be formed and evaluated over a limited number of parameters to define an entire class of these systems.

This gives a fundamentally different result than the normal optical design approach, which is to try to arrive at a single solution that best meets a set of predetermined goals. In that case, whether the end is reached with ray-tracing software (as is most common), or via an analytical solution such as that outlined by Korsch [10], the normal end result is one particular system, or maybe several.

What is described here is a different idea. Instead of aiming for one ultimate system design, the approach is to define a class of systems and evaluate, with sufficient density, points on all necessary parameter axes to completely map the whole possible solution space. Once all possibilities are identified, they can be evaluated against any given set of criteria to see whether something suitable has been found. A general approach could be to start with the simplest possible systems, then increase complexity, if need be, until something attractive is arrived at.

This approach would have been unfeasible in the days prior to electronic digital computers, and it seems fair to say that since computers have become more available and more robust, ray tracing has become the default way to conduct optical design. Ray tracing is very powerful, but it also has limitations. Even with powerful computers, global optimisation is still a demanding activity, and it seems to be impossible to tell whether or not one has reached the true global minimum in relatively simple landscapes. The amount of computation required to arrive at one solution with ray tracing is orders of magnitude greater than that required to trace several paraxial rays and compute derived parameters, as we have shown here. This makes the ray-tracing-based methods for finding solutions ill-suited to dense mapping over useful ranges of parameter space.

On the other hand, the approach described here does just that. It gives a true map of all possibilities within the boundaries of the class under consideration and with relatively little effort. When this survey was first conducted in 2000 [11], it found new types of systems that had not been previously discovered, despite around 60 years of active and growing interest in the field of three-mirror anastigmat design [12].

The particular system class mapped here is that of 'simplest possible reflecting anastigmats'. By this, we mean to consider the set of reflecting anastigmats that produce four non-coincident plates (so we exclude special cases with coincident plates that have already been discussed). As was discussed in section 3.2, this restricts us to:

- two-mirror systems with two aspheric surfaces,
- three-mirror systems with one aspheric surface, and
- four-spherical-mirror systems.

For brevity, we shall refer to any of these cases of 'simplest possible reflecting anastigmats' as 'simplex anastigmats'. Thus the 'simplex three-mirror anastigmat (TMA)' means a three-mirror anastigmat with only one mirror aspheric.

The following description is covered in more detail in the references. Mathematica code is provided in appendix 3, which can be used to reproduce the results described here in part, with the reference material pointing the way to the full method. Using the tools provided in this book, it should be relatively straightforward to replicate the results. In the following, we shall restrict ourselves to an overview of the method.

As was the case with the two-mirror anastigmat, the first step is to define an equation in several parameters, which can then be solved for any combination of input parameters. In the two-mirror case, we could select starting values for r_1 and y_1 without any loss of generality, as the resultant systems could be scaled. Then, for each of the convex and concave primary mirror cases, each value of t_1 gives a unique solution, which may or may not be a practical system.

The two solution sets were one-dimensional curves in two-space. If we consider the object conjugate distance as another variable, then sets of viable two-mirror anastigmats are defined in surfaces in three-space.

In the case of simplex TMAs, we gain an extra dimension in our solution sets and also more solution sets. As we shall see, this gives rise to a considerably greater range of viable geometries than was the case with two mirrors.

In our simplex TMA, the aspheric surface can be on any of the mirrors. We break the solution set into three subsets, depending on which of M1, M2, or M3 is to be aspherised.

Considering the set with M1 aspherised, we take the step of defining the system stop to be on M1. This means that the aspheric term makes no contribution to coma or astigmatism.

We can then solve plate equations as follows:

$$x_{S1} W_{S1} + x_{S2} W_{S2} + x_{S3} W_{S3} = 0 \text{ (coma)}$$

and

$$x_{S1}^2 W_{S1} + x_{S2}^2 W_{S2} + x_{S3}^2 W_{S3} = 0 \text{ (astigmatism)}. \tag{5.7}$$

We can already evaluate the quantities associated with M1, and so for any M1–M2 separation and M2 radius, we can obtain x_{S3} and W_{S3} as follows. First, rearranging:

$$x_{S3} W_{S3} = -(x_{S1} W_{S1} + x_{S2} W_{S2}),$$

$$x_{S3}^2 W_{S3} = -(x_{S1}^2 W_{S1} + x_{S2}^2 W_{S2}),$$

giving

$$x_{S3} = \frac{x_{S3}^2 W_{S3}}{x_{S3} W_{S3}} = \frac{-(x_{S1}^2 W_{S1} + x_{S2}^2 W_{S2})}{-(x_{S1} W_{S1} + x_{S2} W_{S2})}. \tag{5.8}$$

Having obtained x_{S3}, we then obtain W_{S3} from the coma expression:

$$W_{S3} = \frac{-(x_{S1} W_{S1} + x_{S2} W_{S2})}{x_{S3}}. \tag{5.9}$$

In this way, for any chosen values of M1–M2 separation (t_1) and M3 radius (r_2), we can evaluate equations (5.8) and (5.9) and determine both the location of the centre of curvature of M3 (via x_{S3}) and what its spherical aberration contribution must be to produce an anastigmatic system BEFORE we know its radius of curvature or spacing from M2.

We can calculate the M3 radius using an expression for spherical aberration that we did not define in chapter 2 but which is achieved in some simple steps by considering terms defined in figure 2.5 (reproduced below in figure 5.7 for convenience).

From the figure, we can see that:

$$u = \phi - i = \frac{y - \mathrm{P}}{r} = c(y - \mathrm{P}),$$

$$\rightarrow y = \frac{u + c\mathrm{P}}{c}, \text{ and } i = c\mathrm{P}. \tag{5.10}$$

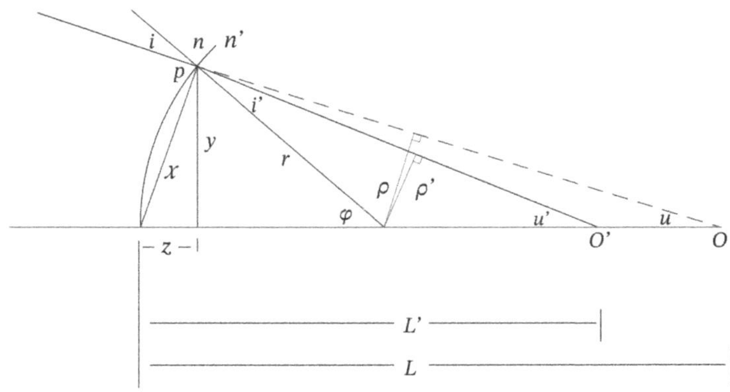

Figure 5.7. Reproduction of figure 2.5 defining quantities in terms of P, which is the length of the perpendicular from the centre of curvature of the surface to the incident ray.

Substituting these into equation (2.28) and expressing in terms relevant to M3 gives:

$$\begin{aligned} W_{S3} &= -\frac{1}{4} n_3 c_3 i_3^2 y_3^2, \\ &= -\frac{1}{4} n_3 c_3 (c_3 \mathrm{P}_3)^2 \left(\frac{u_3 + c_3 \mathrm{P}_3}{c_3} \right)^2, \\ &= -\frac{1}{4} n_3 c_3 \mathrm{P}_3^2 (u_3 + c_3 \mathrm{P}_3)^2, \end{aligned} \tag{5.11}$$

which is cubic in c_3.

We can evaluate P_3 at the tertiary mirror by doing a single ray trace from the location of x_3 in object space (obtained from equations (5.8)), through M1 and M2, to locate the distance l_3 from M2 to the location of the centre of curvature of M3. Then, making a transfer matrix using this value l_3, we can trace a marginal axial ray from object space through mirrors M1 and M2 and the new transfer matrix. The value of y_+ thus obtained is P_3. The mathematical detail can be followed in appendix 3.

It remains to solve the cubic in equations (5.1). Rearranging gives:

$$W_{S3} + \frac{1}{4} n_3 c_3 \mathrm{P}_3^2 (u_3 + c_3 \mathrm{P}_3)^2 = 0. \tag{5.12}$$

If we let:

$$a_0 = \frac{W_{S3}}{\mathrm{P}_3^4},$$

$$a_1 = \frac{u_3^2}{\mathrm{P}_3^2},$$

$$a_2 = \frac{2u_3}{P_3},$$

and using these, produce:

$$Y = 3a_1 - a_2,$$

$$Z = -27a_0 + 9a_1a_2 - a_2^3$$

$$Q = \sqrt{4Y^3 + Z^2},$$

then three expressions can be produced for the explicit solutions to equations (5.1) as follows:

$$\mathrm{sol}_1 = -\frac{a_2}{3} - \frac{\sqrt[3]{2}\,Y}{\sqrt[3]{Z+Q}} + \frac{\sqrt[3]{Z+Q}}{3\sqrt[3]{3}},$$

$$\mathrm{sol}_2 = -\frac{a_2}{3} + \frac{(1 - i\sqrt{3})Y}{\sqrt[3]{4(Z+Q)}} - \frac{(1 + i\sqrt{3}).\,\sqrt[3]{Z+Q}}{6\sqrt[3]{2}},$$

$$\mathrm{sol}_3 = -\frac{a_2}{3} - \frac{(1 - i\sqrt{3})Y}{\sqrt[3]{4(Z+Q)}} + \frac{(1 + i\sqrt{3}).\,\sqrt[3]{Z+Q}}{6\sqrt[3]{2}}. \quad (5.13)$$

Solving these equations gives three different values for c_3, from each of which the remaining system parameters (M2–M3 separation and M3–focal surface distance) can be quickly calculated. Therefore, for a simplex TMA with an aspheric primary mirror, we can say there are three independent solutions for any pair of coordinates (t_1, c_2). These solutions can then be mapped over the (t_1, c_2) plane and their full properties determined for any point.

As with the two-mirror case, there is no guarantee that a solution to these equations, as formulated, is going to be a practical system. In the three-mirror case, we also see many regions that give rise to solutions with imaginary components, which are not physically practical. t_1 and c_2 can solve for the full system at each point, and a few conditional loops can eliminate any solutions with unrealistic properties or undesirable properties from consideration.

For example, figures 5.8–5.14 show solution maps for simplex TMAs with M1 aspherised, subject to the following conditions: (1) the real-to-imaginary component ratio exceeds a chosen cutoff ($\frac{\mathrm{Sol}_{RE}}{\mathrm{Sol}_{IM}} > 10^6$ in this case); (2) air spaces are real, not virtual; (3) y_3 is constrained so that it cannot exceed 1.5 times the M1 semi-diameter.

The plate-diagram approach is not uniquely capable of producing such fully mapped solution spaces. An equivalent approach could be taken using, say, a Seidel sum setup. But this has not yet been done, and in the author's opinion, one of the reasons for this may be complexity. The plate equations lend themselves to intuitively recognising the simplifying steps given above that make this sort of approach possible. For example, the realisation that the centre of curvature of M3 can be found without knowing the M3 radius of curvature is something that becomes obvious when the plate equations are formulated.

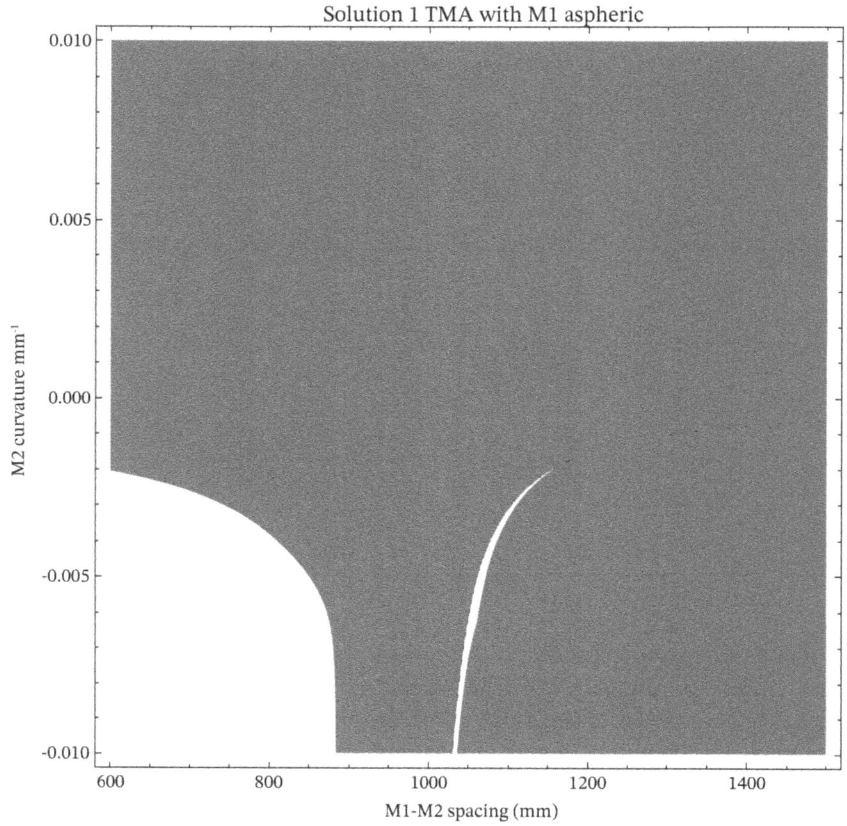

Figure 5.8. Simplex TMA with solution points colour-coded depending on the sign of the system Petzval curvature. White points are positive, black points have negative curvature. The very thin curving white region just after the prime focus actually has a very narrow black region directly abutting on its left-hand side, but it is too narrow to be visible in this image. The curve formed where white and black regions abuts represents a locus of solutions that have a flat field. Cook's flat-field TMA is a point on this curve. Note that all valid solutions here have negative M2 curvature, so they are all of the concave M1/concave M2 type.

Whatever the reason, the fact is that despite the large amount of work done in this field prior to the year 2000, this had not been done. Lacy Cook, a noted American pioneer in the field of TMA design and analysis, discovered several interesting simplex TMAs [13]. These are represented by single points in this parameter-space mapping. Cook also developed a system for categorising TMAs based on the sign of the power of each of the mirrors. For example, '+−+', represents a TMA with positive, negative, and positive powers in M1, M2, and M3, respectively. Cook pointed out in 1992 that 'so far a '++−' system has not been found' [14].

This survey found such a system. It also identified four families of flat-field TMAs, three of which had not previously been found [12]. Cook had already found one of these; another is the Paul–Rumsey type discussed earlier.

A few example layouts from these solution spaces are presented in figures 5.1–5.15.

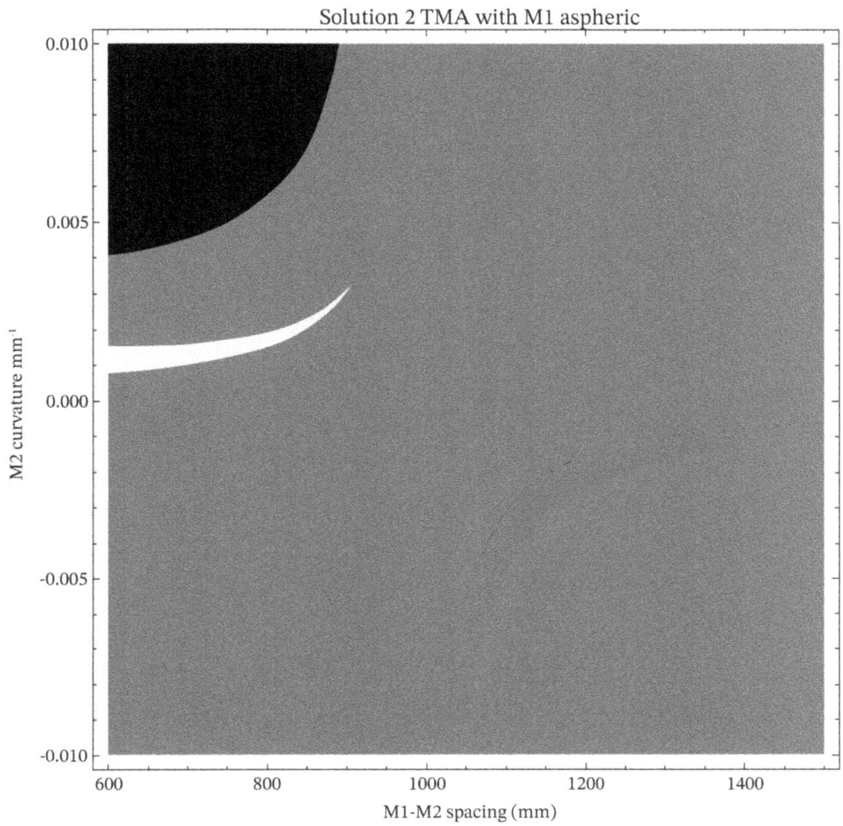

Figure 5.9. Systems corresponding to Solution 2 in equation (5.13). There are no flat-field variants but two distinct regions of valid solutions. All these solutions have convex secondary mirrors.

The reader can use the references given in this survey to obtain more details, including specific design data. In particular, one paper discusses the flat-field families.

The references also discuss how solutions that are 100% self-obstructing, when used in the rotationally symmetrical mode shown here, can become completely unobstructed Schiefspiegler designs when used with some combination of off-axis pupil and field centre. A third means of reducing self-obstruction, namely tilting components so that axial symmetry is lost, is discussed only in the concluding comments of this edition of the book, but some combination of these three possibilities considerably broadens the range of viable solutions that are excluded by rotational symmetry.

The Mathematica code in appendix 3 can be used to produce any number of thousands of these solutions, filtered any way that suits. In the current code, some 'running data' is used to keep track of solutions as the matrix of solutions is being calculated. A sample of this is shown in figure 5.15. This data is from Solution 3, and any of these parameter rows can be used to produce an anastigmat.

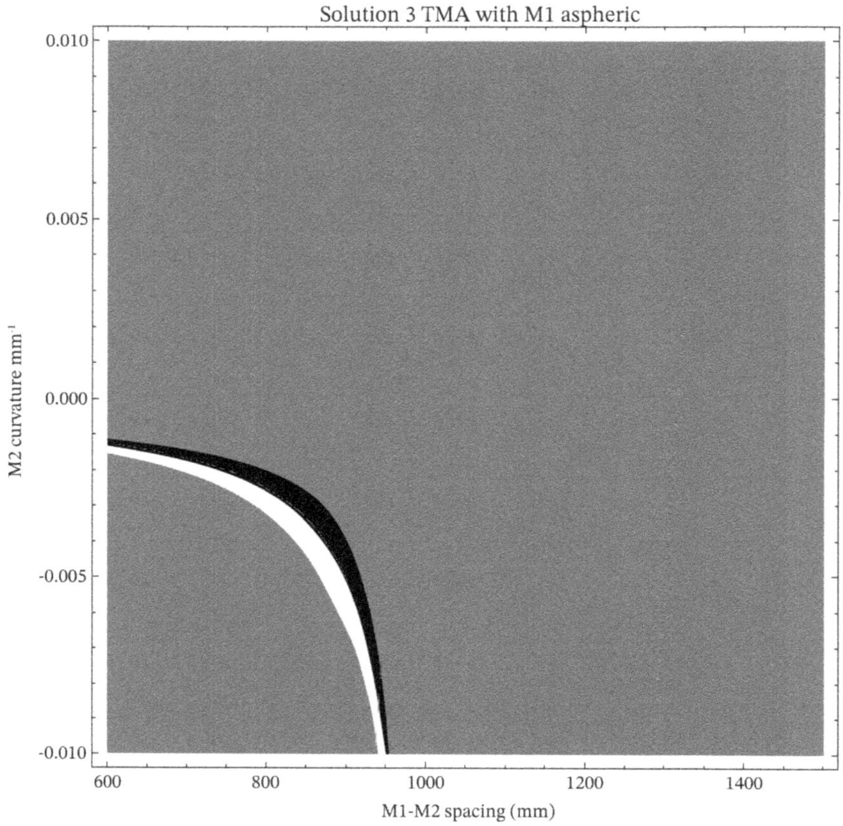

Figure 5.10. In the final solution set, we see that another flat-field type exists. In this case, all variants again have convex secondary mirrors. The classic Paul TMA, the Paul–Rumsey, and Willstrop's 'Mersenne–Schmidt' variants of the Paul are all contained in this solution set. There are no valid solutions beyond the prime focus (the chosen 2000 mm primary mirror has its principal focus at 1000 mm).

The complete solution sets include three sets each for the simplex anastigmats with only M2 aspherised or only M3 aspherised, respectively. Note that this is only for the infinite conjugate. Equivalent sets can be produced for any finite-conjugate-distance pair, so the complete solution spaces, counting the object conjugate distance as a variable, are three-dimensional. Note also that these are only the concave primary mirror versions. The Schwarzschild flat-field two-mirror anastigmat has been widely used since its invention, and that has a convex primary mirror. The same can be said for the concentric spherical-mirror microscope objective. A large number of potentially interesting TMA microscope objectives can be found by this method, but this work has not yet been done.

These solution sets represent something different from all the previous work on TMAs. While there are hundreds if not thousands of papers describing this TMA or that, i.e. single solutions, this work defines a whole class. Every possible solution is revealed, and the natural mathematical 'genetics' underlying each solution set is

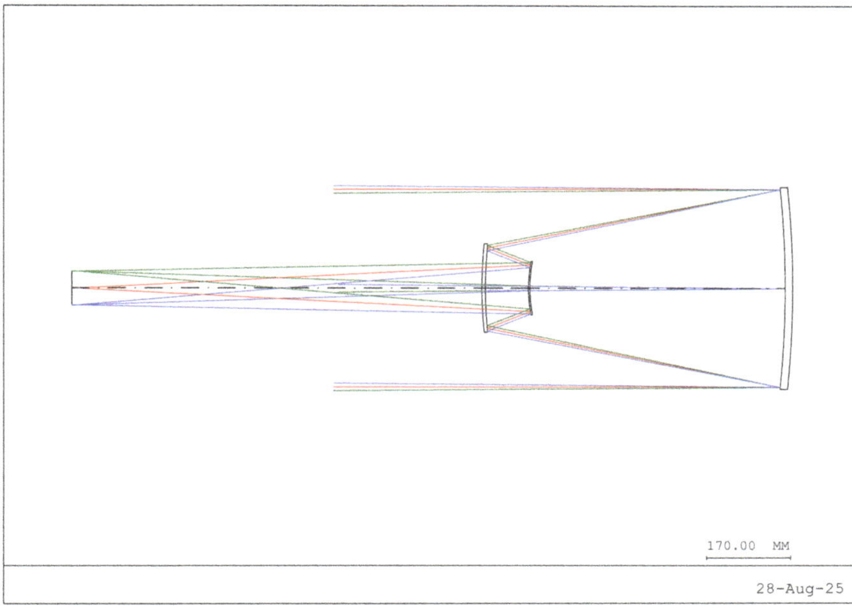

Figure 5.11. Concave secondary type with a strong telephoto effect and 60% linear obscuration for a 0.7 degree field. This is from the Solution 2 set. This is of the + + - type that Cook pointed out had not been found in 1992. Created with Code V.

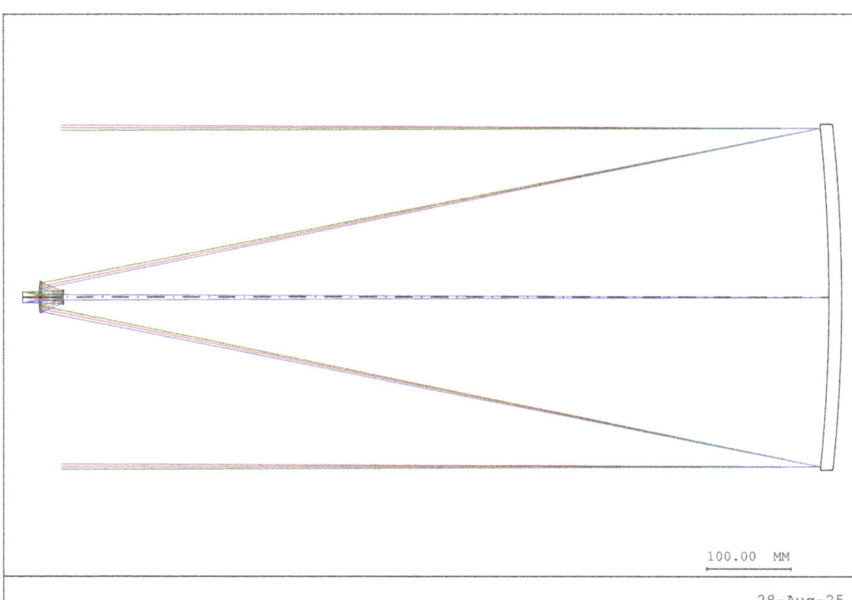

Figure 5.12. A small, spherical-mirror corrector for a mild hyperboloid primary mirror. This anastigmat has a 2x telephoto effect and can be produced at 50% linear obstruction for fields under 1 degree. It is again of the + + - type discussed by Cook. Created with Code V.

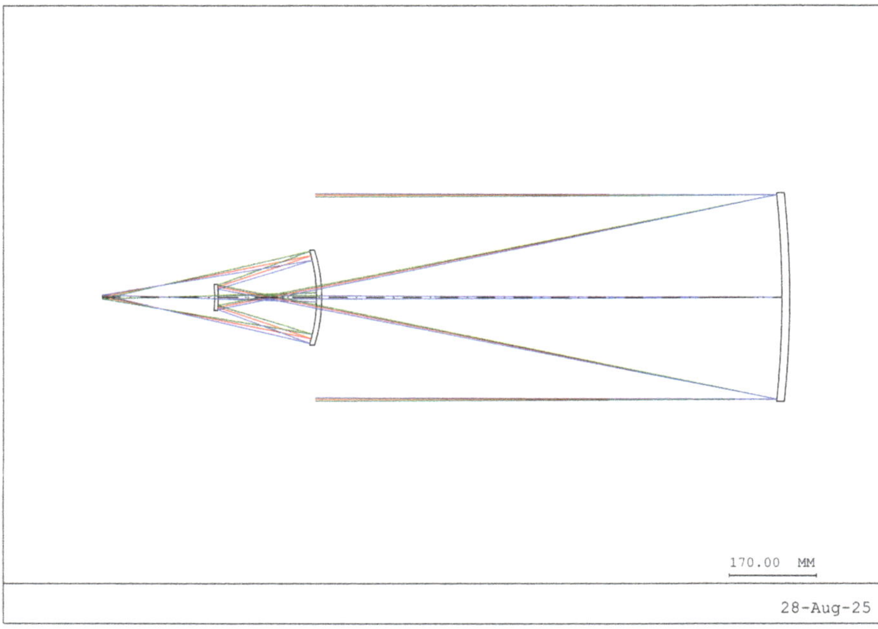

Figure 5.13. A flat-field anastigmat of the type discovered by Cook from the Solution 1 set. It can be made practical by adding a fold mirror near the prime focus. Created with Code V.

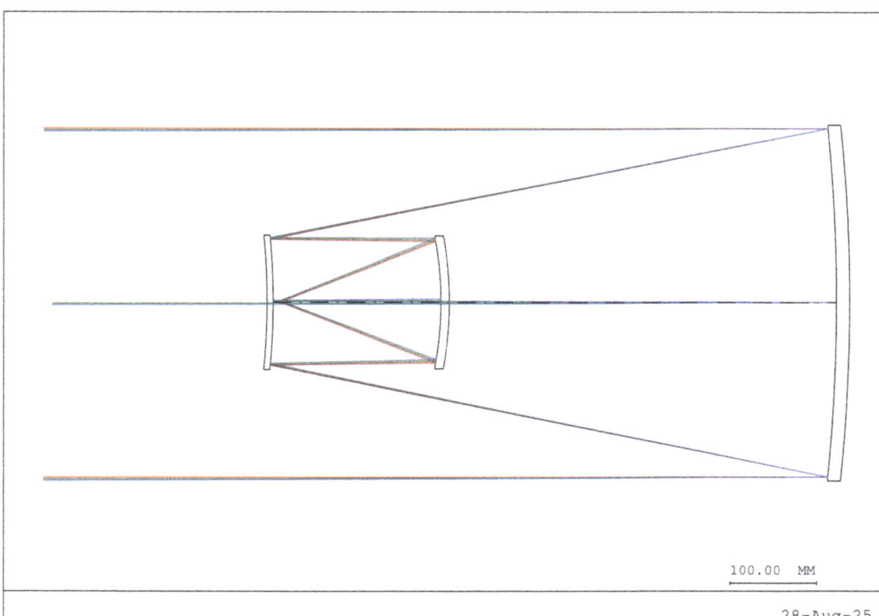

Figure 5.14. An example from the Solution 3 set, showing that a more accessible focus than that of the Paul or its variants is also possible. Created with Code V.

r1	k_1	t1	r2	t2	r3	t3	y3	F/#
-2000	-1.55954	-731.5	-597.015	222.601	-356.24	-166.041	49.2246	-1.68656

Ws1	Ws2	Ws3	Xs1	xS2	xS3	spha	coma	asti	Sol_{Raw3}
0.05	-0.0146252	0.0426021	-2000	-4044.	959.002	-6.93889×10^{-18}	0.	0.	$-0.0028071 - 2.16417\times10^{-19}\, i$

r1	k_1	t1	r2	t2	r3	t3	y3	F/#
-2000	-1.67057	-734.	-597.015	198.014	-340.18	-158.109	48.8872	-1.61708

Ws1	Ws2	Ws3	Xs1	xS2	xS3	spha	coma	asti	Sol_{Raw3}
0.05	-0.0145735	0.0481022	-2000	-4021.01	.860.666	-6.93889×10^{-18}	0.	0.	$-0.00283962 - 4.33257\times10^{-19}\, i$

r1	k_1	t1	r2	t2	r3	t3	y3	F/#
-2000	-1.80651	-736.5	-597.015	173.802	-324.191	-150.34	48.6235	-1.54596

Ws1	Ws2	Ws3	Xs1	xS2	xS3	spha	coma	asti	Sol_{Raw3}
0.05	-0.0145176	0.0548434	-2000	-3998.37	764.963	0.	0.	0.	$-0.0030846 - 4.32834\times10^{-19}\, i$

r1	k_1	t1	r2	t2	r3	t3	y3	F/#
-2000	-1.97643	-739.	-597.015	149.794	-308.118	-142.656	48.4357	-1.47264

Ws1	Ws2	Ws3	Xs1	xS2	xS3	spha	coma	asti	Sol_{Raw3}
0.05	-0.0144578	0.0632795	-2000	-3976.06	671.859	-1.38778×10^{-17}	7.10543×10^{-15}	0.	$-0.00324551 - 2.16417\times10^{-19}\, i$

Figure 5.15. Running output used for engineering checks in the current software.

naturally grouped according to which of the three cubic analytical expressions was used to produce the given set. Unlike a TMA study done via ray tracing, or even via analytical work that, for example, solves Korsch's equations to give a particular anastigmat, this survey of parameter space is the 'final word' for the class of TMAs under investigation. There are no more than those revealed here, and no more can ever be found.

Another extension of this work, leading to larger numbers of solutions with desirable first-order characteristics, would preselect more of the mirror parameters and make one or both of the spherical mirrors aspheric. An equivalent approach would then map out those spaces. This is described in more detail in the following section, where we consider four-mirror simplex anastigmats.

5.5 Four-mirror systems

In the early 2000s, the simplex anastigmat survey was extended to four-mirror systems. This work was first reported in three SPIE conference papers, subsequently expanded in *Optical Engineering* [15–17], and later bound as a University of Canterbury PhD thesis awarded in 2008 [18].

The first of the *Optical Engineering* papers, 'Four-mirror anastigmats Part 1: a complete solution set for four-mirror telescopic systems', reports the basic method and results, which will not be reproduced here.

It is interesting to consider the approach taken to set up the necessary equations. With four spherical mirrors, the four plates lie at the images of each of these mirrors, centres of curvature in object space. Using a trick similar to that used with three mirrors, we place the system aperture stop at the centre of curvature of the 4th mirror, reducing the plate equations for coma and astigmatism as we did in the three-mirror case by placing the stop on M1 and cancelling the aspheric plate.

In this case, the physical location of the stop was determined. In the four-mirror case, the ultimate location of the centre of curvature image from M4 is 'to be decided' as we have yet to define values for the preceding elements of the system. To accommodate this, a term is introduced, ε, representing the distance of the eventual location of the M4 circle of confusion image in object space from the pole of M1. This offset term is included with the other x_i, so that:

$$x_1 = r_1 \rightarrow x_1 = r_1 - \varepsilon, \text{ say.}$$

A cubic equation can then be produced for the M3 radius of curvature, giving three solutions as in the case described in the previous section. In this case, we do not have an explicit value but rather one defined in terms of ε. Finally, the same process can be repeated to determine the M4 radius, and again a cubic yields three solutions. There are now nine solutions in total for every initial choice in a three-parameter space defined by ε, t_1, and c_2.

The referenced work describes the resultant novel four-spherical-mirror anastigmats, which represent the only possibilities within the evaluated parameter space.

Self-obscuration becomes an increasing problem in n-mirror systems as n increases. There are a large number of unexplored anastigmatic Schiefspiegler solutions within that data set that the author (and no-one else to the author's knowledge) has yet investigated. These are of interest, as none of them require an aspheric element.

Another way to use this design approach is to constrain the first-order properties more tightly, providing a system with good obstruction, a desired f-ratio, etc. and to add aspheres to provide the necessary aberration correction. The second OE paper referenced gives examples of this in the four-mirror case.

In this approach, a first-order layout is initially produced exactly as desired. The system sums for spherical aberration $\left(\sum_{i=1}^{4} W_{Si}\right)$, coma $\left(\sum_{i=1}^{4} x_{Si} W_{Si}\right)$, and astigmatism $\left(\sum_{i=1}^{4} x_{Si}^2 W_{Si}\right)$ are then found using the defined radii, air spaces, and associated calculated quantities.

A plate system of equations can then be produced as follows:

$$W_{A1} + W_{A2} + W_{A3} = -\sum_{i=1}^{4} W_{Si},$$

$$x_{A1} W_{A1} + x_{A2} W_{A2} + x_{A3} W_{A3} = -\sum_{i=1}^{4} x_{Si} W_{Si},$$

$$x_{A1}^2 W_{A1} + x_{A2}^2 W_{A2} + x_{A3}^2 W_{A3} = -\sum_{i=1}^{4} x_{Si}^2 W_{Si}.$$

These equations can be solved to give the conic constants of each of the three mirrors. The next step is to systematically cancel aspheres, freeing up the most convenient radii and air spaces as necessary to allow enough variables to balance a plate diagram, followed by a search of the resultant parameter spaces.

As the reference shows, solutions can be found for the example given that very closely match the original first-order properties but require only two or even one aspheric surface. This approach allows for a systematic reduction of complexity while generating a complete solution set at each step.

It has been demonstrated that, in this way, systems can be obtained that both practically match specific requirements and have the minimum possible complexity.

References

[1] Aldis H L 1900 On the construction of photographic objectives *Photogr. J.* **30 June** 291–9
[2] Schwarzschild K 1905 Untersuchungen zur geometrischen Optik II. Theorie der Spiegelsysteme und der astronomischen Instrumente *Math. Ann.* **61** 504–43
[3] Burch C R 1943 On aspheric anastigmatic systems *Proc. Phys. Soc.* **55** 433–44
[4] Couder A 1926 Sur un type nouveau de télescope photographique *C. R.Acad. Sci.* **182** 418–20
[5] Erdös P 1959 Mirror anastigmat with two concentric spherical surfaces *J. Opt. Soc. Am.* **49** 877–86
[6] Wynne C G 1969 Two-mirror anastigmats *J. Opt. Soc. Am.* **59** 572–8
[7] Shafer D R 2005 Some odd and interesting monocentric designs *Proc. of SPIE 5865: Tribute to Warren J. Smith: A Legacy in Lens Design and Optical Engineering* vol 586508 (SPIE) 5
[8] Sasian J M 1990 Design of a Schwarzschild flat-field, anastigmatic, unobstructed, wide-field telescope *Opt. Eng.* **29** 1–5
[9] Kavanagh J 1953 *U. S. Patent* 2,664,026. Dec. 29
[10] Korsch D 1991 *Reflective Optics* (San Diego: Academic)
[11] Rakich A 2000 A complete survey of three-mirror anastigmatic reflecting telescope systems with one aspheric surface *MSc thesis* University of Canterbury, Christchurch, New Zealand
[12] Rakich A 2002 Four families of flat-field three-mirror anastigmatic telescopes with only one mirror aspherized (SPIE) 4768–05 Proceedings of 4768: Novel Optical Systems Design and Optimization VSPIE
[13] Cook L G 1987 Wide field of view three–mirror anastigmat (TMA) employing spherical secondary and tertiary mirrors *Proc. SPIE* **766** 158–62
[14] Cook L G 1992 The last three–mirror anastigmat? *Proc. SPIE* **10263** 102630G
[15] Rakich A 2007 Four-mirror anastigmats, part 1: a complete solution set for all-spherical telescopic systems *Opt. Eng.* **46** 103001
[16] Rakich A 2008 Four-mirror anastigmats, part 2: systems with useful first-order layouts and minimum complexity *Opt. Eng.* **47** 033002
[17] Rakich A 2008 Four-mirror anastigmats, part 3: all-spherical systems with elements larger than the entrance pupil *Opt. Eng.* **47** 033003
[18] Rakich A 2008 Simple four-mirror anastigmatic systems with at least one infinite conjugate *PhD thesis* University of Canterbury, Christchurch, New Zealand

Chapter 6

Conclusions

We have now looked at the full story of the plate diagram. The material in this book references every known occurrence in the published literature where the plate diagram is described or used. Considering that we are spanning a period of 185 years, from Petzval to the present day, there is really not much there.

The reader can judge for themselves whether the insights discussed and novel designs uncovered with a relatively small amount of work using this technique justify taking it further.

As mentioned at the outset, this book is not advocating for analytical methods of optical design to replace ray-tracing tools. That would be naïve. Rather, the point is that analytical techniques applied to optical design, and in particular the plate-diagram approach, offer new strengths in areas in which ray tracing alone is relatively weak.

The first is personal insight, and for that, the plate-diagram method seems to the author to be best suited by far for that purpose. The exact same third-order results are arrived at by solving Seidel sums or via Korsch [1] or by taking an Eikonal approach, following, say, Schwarzschild [2] or Buchdahl [3]. But only the plate diagram makes such an immediately visible juxtaposition of the minimum necessary set of information required to see the third-order system. For 50 years, the optical design community missed insights into the asymmetric Offner and, in particular, into the broad class of perfect instruments that can be described as 'plate-corrected unit-magnification afocal relays'. The revelation of these insights through plate-diagram investigation provides strong evidence that only the plate diagram offers such an immediately visible juxtaposition of the minimum necessary set of information required to see the third-order system.

Second, as the last chapters demonstrate, analytical techniques naturally lend themselves to the sort of global exploration of the available solution space that ray tracing cannot practically mimic. It would seem that much more work could be done in this area to develop software tools to streamline this approach. Again, the plate

doi:10.1088/978-0-7503-3099-2ch6

diagram has allowed some nonintuitive simplifications to be made to systems of three and four mirrors, which have enabled the types of surveys described in chapter 5. Ray Wilson described this approach as 'the definitive triumph of third-order theory as applied to setting up telescope solutions', and it has not been replicated by any other approach to date [4].

As a pedagogical tool, the plate diagram is probably unmatched in its ability to rapidly and elegantly convey the 'meaning behind the math' to students. The mechanical lever analogies and four-plate theorem are vivid and compelling insights that aid both comprehension and retention. Students who become experts with the plate diagram develop intuitive understandings that often take experts years to cultivate.

Finally, an interesting point to consider is that this whole process of constructing a plate diagram, which is perhaps a little laborious to work through manually but can be done very quickly with practice, can also be completely automated and developed as part of an automatic lens design program. Virtually any system of any complexity can be broken down into a combination of fixed elements and a minimum set of plate elements to achieve a certain performance, for example, flat-field astigmatism. This can be used as the 'atomic basis' for applying this technique to optical designs of arbitrary complexity.

There is much more work to be done in this area, too, making it a rich area of exploration for any students who 'get the bug'. Off-axis systems [5], freeform design [6], and alternative approaches to nodal aberration theory [7] have all been started and could be developed much further. Along with these, new areas of investigation, such as the addition of high-order aberration terms, further work on refracting systems, and integration with modern optical design software could lead to the development of more powerful design tools for new generations of optical designers.

References

[1] Korsch D 1991 *Reflective Optics* (San Diego, CA: Academic)

[2] Schwarzschild K 1905 Untersuchungen zur geometrischen Optik II. Theorie der Spiegelsysteme und der astronomischen Instrumente *Math. Ann.* **61** 504–43

[3] Buchdahl H A 1960 *An introduction to Hamiltonian optics* (Cambridge: Cambridge University Press)

[4] Wilson R N 2007 *Reflecting Telescope Optics I: Basic Design Theory and its Historical Development* 2nd corr. edn (Berlin: Springer) 232

[5] Simon J M 1993 A new high-resolution monochromator with spherical mirrors *Proc. SPIE* **1983** 174–6

[6] Fuerschbach K, Rolland J P and Thompson K P 2012 Extending nodal aberration theory to include mount-induced aberrations with application to freeform surfaces *Opt. Express* **20** 20140–53

[7] Rakich A 2010 Calculation of third-order misalignment aberrations with the optical plate diagram *Proc. of SPIE 7652: Int. Optical Design Conf. 2010* (Jackson Hole, Wyoming, USA: SPIE) 765230

Appendix A

A.1 Appendix 1: Rumsey-like TMA from section 3.5

```
n0 = 1;
n1 = -1;
n2 = 1;
n3 = -1;
y1 = 500;
r1 = -2450;
xs1 = r1; (*stop on M1*);
r2 = -1108.964;
r3 = -2026.016;
t1 = -735.088;
t2 = 735.088;
AA = {{1, 0}, {(n1 - n0) / r1, 1}};
BB = {{1, -t1 / n1}, {0, 1}};
CC = {{1, 0}, {(n2 - n1) / r2, 1}};
DD = {{1, -t2 / n2}, {0, 1}};
EE = {{1, 0}, {(n3 - n2) / r3, 0}};
(*calculate WS1*)
WS1 = -1 / 4. * 1 / r1^3 * y1^4; (*calculate u'2, y2, i2 and from that WS2*);
   uy1 = {0, 500};
   y2 = (uy1.AA.BB)[[2]];
   uprime2 = n1 * (uy1.AA.BB)[[1]];
   phi2 = y2 / r2;
   i2 = phi2 - uprime2;
   WS2 = -1 / 4 * n1 * 1 / r2 * i2^2 * y2^2; (*calculate u'3, y3, i3 and from that WS3*);
   y3 = (uy1.AA.BB.CC.DD)[[2]];
   uprime3 = n2 * (uy1.AA.BB.CC.DD)[[1]];
   phi3 = y3 / r3;
   i3 = phi3 - uprime3;
   WS3 = -1 / 4 * n2 * 1 / r3 * i3^2 * y3^2; (*Check system sum with Zemax SPHA*); (*Step 2, calculate xs2 and xs3. In this step,
  use the quantities defined below and the INVERSE of the correct combination of ray trace matrices defined above. Now we are reverse tracing so start with a negative
   sign of the distance to teh respective mirrors*); (*xs2prime is the distance of M2 CoC to M1. As the initial direction of light travel is left to right, n is - here*);
 xs2prime = -(t1 + r2); (*xs2prime is the distance of M3 CoC to M2. As the initial direction of light travel is again left to right, because the CoC of M3 is a virtual object,
 behind the M2 surface,n is again - here*);
 xs3doubleprime = -(t2 + r3); (*first calculate xs2, starting at an on-axis point with some arbitrary u*);
 uyobjspcxs2 = {-.1, 0}.{{1, -xs2prime / n0}, {0, 1}}.{{1, 0}, {-2 / r1, 1}};
 xs2 = uyobjspcxs2[[2]] / (n1 * uyobjspcxs2[[1]]); (*next calculate xs3, which is xs3doubleprime, imaged through first M2, then M1*);
 uyobjspcxs3 = {-.1, 0}.{{1, -xs3doubleprime / n0}, {0, 1}}.{{1, 0}, {-2 / r2, 1}}.{{1, -(-t1) / -1}, {0, 1}}.{{1, 0}, {2 / r1, 1}};
 xs3 = uyobjspcxs3[[2]] / (n2 * uyobjspcxs3[[1]]); (*Now step 3 need to calculate images of M2 and M3 vertices in object space giving xA2 and xA3*); (*first calculate xA2*);
```


doi:10.1088/978-0-7503-3099-2ch7

```
uyobjspcxa2 = {-.1, 0}.{{1, -(-t1)/1}, {0, 1}}.{{1, 0}, {-2/r1, 1}};
xA2 = uyobjspcxa2[[2]] / (-1 * uyobjspcxa2[[1]]); (*now calculate xA3*) ;
uyobjspcxa3 = {-.1, 0}.{{1, -(-t2)/1}, {0, 1}}.{{1, 0}, {-2/r2, 1}}.{{1, -(-t1)/-1}, {0, 1}}.{{1, 0}, {2/r1, 1}};
xA3 = uyobjspcxa3[[2]] / (1 * uyobjspcxa3[[1]]); (*solve for conicoid plates*) ;
Solve[{WS1 + WS2 + WS3 + Wa1 + Wa2 + Wa3 == 0 && WS1 * xs1 + WS2 * xs2 + WS3 * xs3 + Wa2 * xA2 + Wa3 * xA3 == 0 && WS1 * xs1^2 + WS2 * xs2^2 + WS3 * xs3^2 + Wa2 * xA2^2 + Wa3 * xA3^2 == 0}, {Wa1, Wa2, Wa3}]
{{Wa1 -> -1.51119, Wa2 -> 0.940526, Wa3 -> -0.0797569}}

WA1 = Wa1 //. {{Wa1 -> -1.5111926959142354`, Wa2 -> 0.9405257088642336`, Wa3 -> -0.0797568942968977`}}[[1]];
WA2 = Wa2 //. {{Wa1 -> -1.5111926959142354`, Wa2 -> 0.9405257088642336`, Wa3 -> -0.0797568942968977`}}[[1]];
WA3 = Wa3 //. {{Wa1 -> -1.5111926959142354`, Wa2 -> 0.9405257088642336`, Wa3 -> -0.0797568942968977`}}[[1]];

(*solve for k1, k2, k3*)

k1 = NumberForm[WA1 / (-.25 * n0 / r1^3 * y1^4), 10]
NumberForm=
-1.422322476

k2 = NumberForm[WA2 / (-.25 * n1 / r2^3 * y2^4), 10]
NumberForm=
-3.209039294

k3 = NumberForm[WA3 / (-.25 * n2 / r3^3 * y3^4), 10]
NumberForm=
-3.577356232

WA1 =.; WA2 =.; WA3 =.;

TableForm[{{"MirrorPlate:Plate Quantity", "xSi", "WSi", "xAi", "WAi"}, {"M1", xs1, WS1, 0, WA1}, {"M2", xs2, WS2, xA2, WA2}, {"M3", xs3, WS3, xA3, WA3}}]
TableForm=
```

MirrorPlate:Plate Quantity	xSi	WSi	xAi	WAi
M1	-2450	1.06248	0	-1.51119
M2	-3649.07	-0.467965	1838.05	0.940526
M3	-4338.02	0.0559068	7407.07	-0.0797569

A.2 Appendix 2: Two-mirror design from section 5.1

```
uy0 = {0, y1};
n0 = 1;
n1 = -1;
n2 = 1;
A = {{1, 0}, {-2/r1, 1}};
B = {{1, -t1/n1}, {0, 1}};
CC = {{1, 0}, {(n2 - n1)/r2, 1}};
y2 = (uy0.A.B)[[2]];
phi2 = y2/r2;
u1 = (uy0.A)[[1]] * -1;
i2 = phi2 - u1;
tm2vert = {{1, -(-t1)}, {0, 1}};
x1s = r1;
x1a = 0;
x2a = ({n0 * uone, 0}.tm2vert.A)[[2]] / (-({n0 * uone, 0}.tm2vert.A)[[1]]);
tm2coc = {{1, -(-t1 - r2)}, {0, 1}};
x2s = ({n0 * uone, 0}.tm2coc.A)[[2]] / (-({n0 * uone, 0}.tm2coc.A)[[1]]);
W1s = -1/4 * y1^4 / r1^3;
W1a = k1 * W1s;
W2s = -1/4 * n1 / r2 * i2^2 * y2^2;
W2a = -1/4 * n1 * k2 * y2^4 / r2^3;
L = uy0.A.B.CC;
LL = L[[2]] / L[[1]]; Solve[{W1s * (1 + k1) + W2s + W2a == 0 && x1s * W1s + x2s * W2s + x2a * W2a == 0 && x1s^2 * W1s + x2s^2 * W2s + x2a^2 * W2a == 0},
 {r2, k1, k2}]
```

$$\left\{\{r2 \to r1 - 2\,t1,\ k1 \to -1,\ k2 \to -1\},\ \left\{r2 \to \frac{-r1\,t1 + 2\,t1^2}{r1 - t1},\ k1 \to \frac{-r1^3 + 2\,r1^2\,t1 - t1^3}{t1^3},\ k2 \to \frac{-r1^2\,t1 + r1\,t1^2 + t1^3}{(r1 - t1)^3}\right\}\right\}$$

```
Plot[Evaluate@{r2 / 1000, k1, k2, (1 / r1 - 1 / r2) * 1000}, {t1, -2150, 2150}, PlotRange → {-4, 4}, PlotStyle → {Blue, Red, Darker[Green], Purple},
 PlotLegends → Placed[LineLegend[{Blue, Red, Darker[Green], Purple}, {"r2/1000", "k1", "k2", "(1/r1 - 1/r2)*1000"}], Right],
 Frame → True, FrameLabel → {Style["t1 (mm)", 14, Bold], None}, ImageSize → Large,
 PlotLabel → Style["Conic constants, scaled R2 and curvature sum for concave M1 2-mirror anastigmats", 14, Bold]]
```

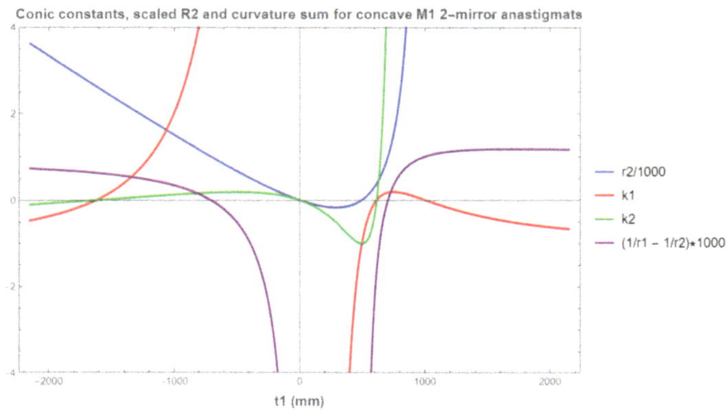

A.3 Appendix 3. TMA survey from section 5.2

```
r₁ = -2000;
c₁ = 1/r₁;
y₁ = 200;
y1 = 200;
n₁ = 1;
n₂ = -1;
n₃ = 1;
n₄ = -1;
n₅ = 1;
t₁₀ = -599;
t₁ = t₁₀;
delt₁ = 2.500001;
Δc₂ = .000025;
sys = -1;
i = 1;
c₂ = -0.010000;
r₂ = 1 / c₂;
h = 1;
VALpetz₃ = {};
VALfnumb₃ = {};
VALm3semiD₃ = {};
(*loop here for cycling t₁*) Label[tcyc];
t₁ = t₁ - delt₁;
If[t₁ ≤ -1499.1, Goto[c2cyc]];
n1uy1 = {0, y1};
A = {{1, 0}, {(n₂ - n₁) / r₁, 1}};
m12 = {{1, -t₁ / n₂}, {0, 1}};
CC = {{1, 0}, {(n₃ - n₂) / r₂, 1}};
```

```
(*later determine t2*) m23 = {{1, -t₂ / n₃}, {0, 1}};
EE = {{1, 0}, {(n₄ - n₃) / r₃, 1}};
(*calculate quantities to deterimine W₂*) r₂ = 1 / c₂;
u₂ = (n1uy1.A) [[1]] / n₂;
y₂ = (n1uy1.A.m12) [[2]];
ϕ₂ = y₂ / r₂;
i₂ = ϕ₂ - u₂;
(*Calculate W₁ and W₂*) W_S1 = -.25 n₁ c₁^3 y₁^4;
W_S2 = -.25 n₂ c₂ i₂^2 y₂^2;
(*Allow Entrance pupil to range over object space. Default to sit on M1 vertex, coordinate system origin*) S = 0;
(*Define reverse transfer matrix for M2 vert and M2 Coc*) m21 = {{1, -(-t₁) / n₁}, {0, 1}};
m2coc1 = {{1, -(-t₁ - r₂) / n₁}, {0, 1}};
uy = {u, 0};
(*uy is launch ray from on axis with arbitrary u*) (*calculate x_1s, x_1A, x_2s, x_2A*) x_S1 = r₁ - S;
x_A1 = 0 - S;
x_S2 = (uy.m2coc1.A) [[2]] / ((uy.m2coc1.A) / n₂) [[1]];
x_A2 = (uy.m21.A) [[2]] / ((uy.m21.A) / n₂) [[1]];
(*here's a trick. For M1 conic... The stop is on M1, so no coma or astigmatism from M1 conic. We have M1 and M2 spheres and M1M2 spacing,
we need M3. We can calculate x3 and W3 from Plate Equations, from that get M3 radius and position x_1s W_1s + x_2s W_2s + x_3s W_3s = 0,
and x_1s^2 W_1s + x_2s^2 W_2s + x_3s^2 W_3s = 0 ...*) x_S3 = (x_S1^2 W_S1 + x_S2^2 W_S2) / (x_S1 W_S1 + x_S2 W_S2);
W_S3 = -(x_S1 W_S1 + x_S2 W_S2) / x_S3;
(*Now we calculate the CoC location of M3, and we know its spherical aberration contribution
```

but not its radius

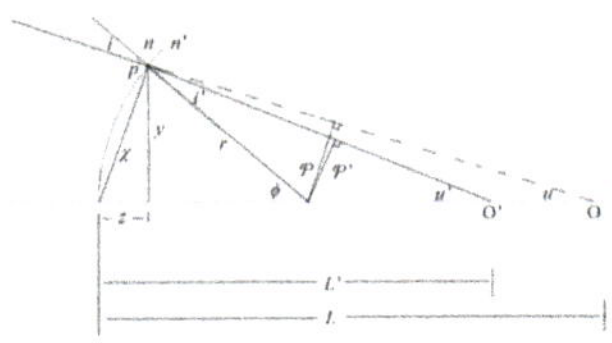

and

$u = \phi - i = \frac{(y - P)}{r} = c(y - P)$

$\Rightarrow y = \frac{(u + cP)}{c}$;

```
W = -1/4 n c i² y⁴ = -1/4 n c (c P)² ((u + c P)/c)² = -1/4 n c P² (u + c P)², which is cubic in c.
   From this we come to W_3s + 1/4 n c P² (u + c P)² = 0, which we can solve analytically for c. First, we need to get P,
which we can calculate by subtracting the calculated distance from M2 to the marginal ray's axis intercept after reflection from M2,
and the distance from M2 of the CoC of M3,
which we obtain by imaging the x3 plate location through M1 and M2. note that x_3s automatically has any S stop shift in
 object space value built in by solving from the other values that had S already in them. We solve for CoC M3 from x_3s by
 imaging through M1 and M2, then dividing the resultant y by the resultant u. first a launch vector and transfer matrix to M1*) uyx3s = {ux3, 0};
Tx3 = {{1, x_S3}, {0, 1}};
(*careful with the minus sign on x₃ as her ex₃ measures from ten stop to the image but reverse tracing we go from image to stop *)
u_x3 = (uyx3s.Tx3.A.m12.CC) [[1]];
y_x3 = (uyx3s.Tx3.A.m12.CC) [[2]];
CoC₃ = FullSimplify[y_x3 / u_x3];
(*Next we calculate the intercept of the marginal axial ray from object space and the axis,
giving us O from the figure. From this and CoC M3 we can calculate P, then from that we get a cubic in c.*) y₂ = (n1uy1.A.m12.CC) [[2]];
u₃ = (n1uy1.A.m12.CC) [[1]];
O₃ = FullSimplify[y₂ / u₃];
l₃ = O₃ - CoC₃;
P₃ = u₃ l₃;
(*Concerning the analytical solution to W_3s + 1/4 n c P² (u + c P)² = 0..., using "Numerical Recipes" came up with the following*) a₀ = 4 W_S3 / P₃^4;
a₁ = u₃^2 / P₃^2;
a₂ = 2 u₃ / P₃;
(*Note there was an annoying sign error on the a₂ term in the thesis*) Y = 3 a₁ - a₂^2;
Z = -27 a₀ + 9 a₁ a₂ - 2 a₂^3;
Q = (4 Y^3 + Z^2)^0.5`;
(*Sol_Raw1 = -a2/3 - 2^(1/3) Y / (3 (Z+Q)^(1/3)) + (Z+Q)^(1/3) / (3*2^(1/3));
If[Abs[Im[Sol_Raw1]] > 10^-6, Sol₁ = 0, Goto[sol1process]];
Sol_Raw2 = -a2/3 + (1 + I √3) Y / (3 (4 (Z+Q))^(1/3)) - (1 - I √3) (Z+Q)^(1/3) / (6*2^(1/3));
If[Abs[Im[Sol_Raw2]] > 10^-6, Sol₂ = 0, Goto[sol2process]];
Sol_Raw3 = ...;*) Sol_Raw3 = -a₂/3 + (1 + I √3) Y / (3 (4 (Z + Q))^(1/3)) - (1 - I √3) (Z + Q)^(1/3) / (6 * 2^(1/3));

If[Abs[Re[Sol_Raw3] / Im[Sol_Raw3]] < 10^6, Goto[Imsol]];
Sol₃ = Re[Sol_Raw3];
r3₃ = 1 / Sol₃;
t2₃ = CoC₃ - r3₃;
m22 = {{1, -t2₃ / n₃}, {0, 1}};
EE = {{1, 0}, {-2 / r3₃, 1}};
y₃₃ = (n1uy1.A.m21.CC.m22.EE) [[2]];
u₃₃ = -(n1uy1.A.m21.CC.m22.EE) [[1]];
```

```
t43 = y33 / u33;
F43 = t43 / (2 y33);
If[t23 < 0, Goto[Imsol]];
If[t43 > 0, Goto[Imsol]];
If[Abs[y33] > 300, Goto[Imsol]];
Petzsol3 = 2 * (c1 - c2 + Sol3);
If[Abs[Petzsol3] ≥ .01, Petzsol3 = 0];
VALpetz3 = Append[VALpetz3, Petzsol3];
k1 = (WS1 + WS2 + WS3) * 4 * r1^3 / y1^4;
Print[MatrixForm[{{"r1", "k1", "t1", "r2", "t2", "r3", "t3", "y3", "F/#"}, {r1, k1, t1, r2, t23, 1 / Sol3, t43, y33, F43}}]];
Print[MatrixForm[{{"Ws1", "Ws2", "Ws3", "Xs1", "xS2", "xS3", "spha", "coma", "asti", "SolRaw3"},
   {WS1, WS2, WS3, XS1, XS2, XS3, WS1 * (1 + k1) + WS2 + WS3, XS1 * WS1 + XS2 * WS2 + XS3 * WS3, XS1^2 * WS1 + XS2^2 * WS2 + XS3^2 * WS3, SolRaw3}}]];
Goto[tcyc];
Label[Imsol];
VALpetz3 = Append[VALpetz3, 0];
Goto[tcyc];
Label[c2cyc];
c2 = c2 + Δc2;
t1 = t10;
If[Abs[c2] > .01, Goto[end], Goto[tcyc]];
Label[end]
```

```
Label[end]
```

r1	k_1	t1	r2	t2	r3	t3	y3	F/#
-2000	-0.97777	-934.	-100.	2352.24	-2032.9	-1686.46	163.742	-5.14974

Ws1	Ws2	Ws3	Xs1	xS2	xS3	spha	coma	asti	Sol_{Raw3}
0.05	-0.00201422	0.000902712	-2000	-30411.6	42919.7	-5.63785×10^{-18}	0.	-2.32831×10^{-10}	$-0.000491908-1.89735\times10^{-19}\,i$

r1	k_1	t1	r2	t2	r3	t3	y3	F/#
-2000	-0.97685	-936.5	-100.	974.483	-803.255	-601.251	55.3215	-4.60224

Ws1	Ws2	Ws3	Xs1	xS2	xS3	spha	coma	asti	Sol_{Raw3}
0.05	-0.00214879	0.000991315	-2000	-28397.2	39321.9	-8.67362×10^{-19}	7.10543×10^{-15}	2.32831×10^{-10}	$-0.00124493-6.23416\times10^{-19}\,i$

r1	k_1	t1	r2	t2	r3	t3	y3	F/#
-2000	-0.977117	-939.	-100.	530.332	-428.254	-291.384	35.5343	-4.10004

Ws1	Ws2	Ws3	Xs1	xS2	xS3	spha	coma	asti	Sol_{Raw3}
0.05	-0.00226386	0.00111973	-2000	-26640.9	35444.8	8.67362×10^{-19}	0.	2.32831×10^{-10}	$-0.00233506-1.89735\times10^{-19}\,i$

r1	k_1	t1	r2	t2	r3	t3	y3	F/#
-2000	-0.978802	-941.5	-100.	330.492	-268.391	-167.518	22.9365	-3.65178

Ws1	Ws2	Ws3	Xs1	xS2	xS3	spha	coma	asti	Sol_{Raw3}
0.05	-0.00235759	0.00129768	-2000	-25096.3	31466.3	1.30104×10^{-18}	0.	0.	$-0.00372591-8.60585\times10^{-19}\,i$

r1	k_1	t1	r2	t2	r3	t3	y3	F/#
-2000	-0.982202	-944.	-100.	222.449	-186.346	-107.74	16.5386	-3.25723

Ws1	Ws2	Ws3	Xs1	xS2	xS3	spha	coma	asti	Sol_{Raw3}
0.05	-0.00242852	0.00153862	-2000	-23727.2	27542.8	-8.67362×10^{-19}	0.	-2.32831×10^{-10}	$-0.00536635-1.29427\times10^{-18}\,i$

r1	k_1	t1	r2	t2	r3	t3	y3	F/#
-2000	-0.987711	-946.5	-100.	156.751	-138.87	-75.096	12.8944	-2.91197

Ws1	Ws2	Ws3	Xs1	xS2	xS3	spha	coma	asti	Sol_{Raw3}
0.05	-0.00247556	0.0018611	-2000	-22505.3	23796.	2.1684×10^{-19}	0.	0.	$-0.00720099-8.65668\times10^{-19}\,i$

r1	k_1	t1	r2	t2	r3	t3	y3	F/#
-2000	-0.995853	-949.	-100.	113.382	-108.948	-55.6115	10.6534	-2.61003

www.ingramcontent.com/pod-product-compliance
Ingram Content Group UK Ltd.
Pitfield, Milton Keynes, MK11 3LW, UK
UKHW051852100926
8556IPUK00005B/580

9 780750 33